Lecture Notes in Mathematics

Edited by A. Dold and B. Eckmann

575

K-Theory and Operator Algebras

Proceedings of a Conference held at the University of Georgia in Athens, Georgia, April 21–25, 1975

Edited by
B. B. Morrel and I. M. Singer

Springer-Verlag
Berlin · Heidelberg · New York 1977

Editors
Bernard B. Morrel
Department of Mathematics
Swain Hall East, Indiana University
Bloomington, IN 47401/USA

I. M. Singer
Department of Mathematics
Massachusetts Institute of Technology
Cambridge, MA 02139/USA

Library of Congress Cataloging in Publication Data

Conference on K-Theory and Operator Algebras, University of Georgia, 1975.
K-theory and operator algebras.

(Lecture notes in mathematics ; 575)
1. K-theory--Congresses. 2. Operator algebras--Congresses. I. Morrel, Bernard B. II. Singer, Isadore Manuel, 1924- III. Title. IV. Series: Lecture notes in mathematics (Berlin) ; 575.
QA3.L28 no. 575 [QA612.33] 510'.8s 77-3050
[514'.23]

AMS Subject Classifications (1970): 18F25, 46-02, 46L05, 46L10, 47-02, 47G05, 55-02, 55B15, 58-02, 58G10, 58G15

ISBN 3-540-08133-X Springer-Verlag Berlin · Heidelberg · New York
ISBN 0-387-08133-X Springer-Verlag New York · Heidelberg · Berlin

Printing and binding: Beltz Offsetdruck, Hemsbach/Bergstr.
2141/3140-543210

Preface

This volume records most of the talks given at the Conference on K-theory and Operator Algebras held at the University of Georgia in Athens, Georgia, April 21-April 25, 1975. The purpose of the conference was to review the known connections between operator theory and K-theory and explore possible new ones. Consequently, some of the papers present historical background, some develop new ideas, some are expository in an attempt to acquaint experts in the one field with recent developments in the other, and some pose new problems which further developments might solve.

We are happy to express our thanks to the National Science Foundation for sponsoring the meeting, to the University of Georgia for providing funds for additional participants, to the Mathematics Department for its gracious hospitality, to Ms. Teddy Schultz, Ms. Carol Ledbetter and Ms. Ann Ware for the typing, to the authors for their manuscripts and their patience, and to Springer-Verlag for publishing this volume.

Bernard Morrel
I. M. Singer

CONTENTS

Conference Participants

Joel Anderson
Michael Atiyah
Edward Azoff
Richard Bouldin
John Bunce
Manfred Breuer
Lawrence G. Brown
Richard Carey
Kevin Clancey
D. N. Clark
Lewis Coburn
E. H. Connell
Iain Craw
James Deddens
James Deel
Allen Devinatz
R. G. Douglas
Maurice Dupre
David A. Edwards
John Ernest
Nazih Faour
Peter Fillmore
Robert E. Goad
E. C. Gootman
Paul Halmos
Herbert Halpern
Allen Hatcher
William Helton
Karl Hofmann
John Hollingsworth
Richard Kadison
Jerome Kaminker
H. W. Kim
Dusa MacDuff
Bernard Morrel
Judith Morrel
R. D. Moyer
Paul Muhly
Catherine Olsen
Carl Pearcy
William T. Pelletier
Joel Pincus
Iain Raeburn
William G. Rosen
Shoichiro Sakai
Norberto Salinas
David G. Schaeffer
Claude Schochet
Graeme Segal
I. M. Singer
James Simons
J. G. Stampfli
Robert Szczarba
Masamichi Takesaki
J. L. Taylor
Javier Thayer

Harold Widom

A SURVEY OF K-THEORY

M.F. Atiyah, Oxford

Introduction

In this talk I shall describe the way K-theory enters in various branches of mathematics. I shall follow the historical development and emphasize those aspects of most relevance to this conference. I shall not therefore dwell too much on the more algebraic parts of the subject.

Let me first make some very general remarks on the nature of K-theory. Roughly speaking K-theory may be described as the linear algebra of large matrices, also called stable linear algebra, and it deals primarily with such notions as idempotents (projections) and units (invertible matrices). Its main feature is that it is an abelian theory, despite the non-abelian character of matrices. This stems from the fact that although A and B may not commute, $A \oplus 1$ and $1 \oplus B$ do commute. Thus by increasing the size of our matrices we can, for certain purposes, reduce to an abelian situation.

When we transfer K-theory from one area of mathematics to another certain formal similarities remain. However each area has different problems and techniques, and the success of K-theory hinges on the fact that in many areas it has proved possible to link it up with natural classical problems.

§1. Algebraic Geometry

K-theory was introduced into Algebraic Geometry by Grothendieck, though the preliminary ground-work had been laid by Serre. For an affine variety $V \subset C^n$, with coordinate ring $A(V)$, we consider finitely-generated projective $A(V)$-modules E, i.e. E is a direct summand of a free module $A(V)^N$ and so is given by an idempotent in the $N \times N$ matrix algebra over $A(V)$. The isomorphism classes of such modules form an abelian semigroup under $\oplus$ and $K^0(V)$ is the corresponding abelian group. Thus $K^0(V)$ is the universal group for studying abelian invariants of projective modules.

For a projective variety one can use graded rings and modules or, better still, one can use the geometrical language of vector bundles. Thus we now consider an algebraic vector bundle E over V or equivalently a locally free sheaf over the sheaf of functions of V (the sheaf is given by the sections of E). Again we define $K^0(V)$ although now, since short exact sequences do not split, we impose a stronger equivalence relative so that $K^0(V)$ is universal for abelian invariants a(E) such that $a(E) = a(E') + a(E'')$ whenever

$$0 \to E' \to E \to E'' \to 0$$

is an exact sequence of vector bundles.

Note that the language of vector bundles works also in the affine case since the global sections $\Gamma(E)$ of E form a finitely-generated projection A(V)-module and the correspondence $E \to \Gamma(E)$ is bijective.

The two basic examples, which motivated Grothendieck, of abelian invariants of E are

i) $\chi(E) = \sum_q (-1)^q \dim H^q(Y, E)$

ii) ch(E), the Chern character of E.

Here (i) is the sheaf cohomology Euler characteristic (V assumed projective): it takes values in Z, while ch(E) takes values in the rational cohomology ring of V. Thus we have homomorphisms

$$\chi : K^0(V) \to Z \qquad ch: K^0(V) \to H^*(V,Q)$$

If we use all coherent sheaves instead of locally free ones (in the affine case this just means dropping the restriction that the A(V)-module be projective) we obtain another group denoted by $K_0(V)$. The formal properties of $K^0(V)$ and $K_0(V)$ can now be summarized as follows:

a) $K^0(V)$ is a ring (under $\otimes$) and $K_0(V)$ is a $K^0(V)$-module (note that $\otimes$ does not preserve exact sequences, which is why $K_0(V)$ does not have a ring structure)

b) $K^0(V)$

contravariant functor of V while, for proper maps, $K_0(V)$ is covariant, the map $K_0(V) \to K_0(V')$ generalizing χ (for V' = point, $K_0(V') = Z$).

c) If V is projective non-singular the natural map $K^0(V) \to K_0(V)$ is an isomorphism.

In view of these formal properties we see that there is a close analogy between K^0 and cohomology, with K_0 playing the role of homology, and c) being Poincare duality. This analogy is in fact quite deep and has had a significant influence on the development of K-theory.

§2. Topology

If X is a compact space we can consider topological vector bundles (fibre $\mathbb{C}^n$) over X. These correspond to finitely-generated projective modules over C(X), the ring of continuous complex-valued functions on X. We then form an abelian group $K^0(X)$ as before. Because of the analogy with cohomology described in §2 we then define, for n = 1,2,...

$$K^{-n}(X) = K_{comp}(R^n \times X)$$

where, for a locally compact Y we put

$$K_{comp}(Y) = \text{Ker}\{K(Y \cup \infty) \to K(\infty)\}$$

$Y \cup \infty$ being the one-point compactification of Y. The famous Bott periodicity theorem then asserts that

$$K^{-n}(X) \cong K^{-n-2}(X)$$

and the groups $K^{-n}(X)$ (extended to all $n \in Z$ by periodicity) form a generalized cohomology theory with products. This means they satisfy all the Eilenberg-Steenrod axioms for cohomology except the dimension axiom (note that $K^n(\text{point}) \cong Z$ for n even). If X is embeddable in some R^N (e.g. if X is a polyhedron) Alexander duality asserts that

$$H_q(X) \cong \tilde{H}^{N-q-1}(R^N - X)$$

where $\tilde{H}$ is reduced cohomology, i.e. modulo a base point. We can therefore define groups

$$K_q(X) = \widetilde{K}^{N-q-1}(R^N - X), \text{ where } \widetilde{K} \text{ is reduced K-theory.}$$

Since $R^N - X$ is not compact we must replace it by a compact deformation retract (imposing some mild restriction on the embedding of X in R^N). From this definition it follows that K_* is a K^*-module and we even have a Poincare-duality theorem for suitable manifolds (e.g. complex manifolds).

Although formally satisfactory it is clear that the above definition of K_* is somewhat artificial. It would be nice to have representative objects which were more natural, as we do in algebraic geometry with coherent sheaves. In [2] I made a first tentative step in this direction and the work of Brown-Douglas-Fillmore, which we shall hear about in this conference, provides a very satisfying and natural analytical solution (see §3). It is also interesting to compare the algebraic and topological K-theories of §1 and §2 when they both apply, namely when X = V is a projective algebraic variety. Clearly we have a homomorphism

$$\alpha^0 : K^0_{alg}(X) \to K^0_{top}(X) .$$

Because K_0^{top} is rather artificial it is not so clear how to define

$$\alpha_0 : K_0^{alg}(X) \to K_0^{top}(X) .$$

When X is non-singular one can apply "Poincare-duality" on both sides and the functorial definition of α_0 is closely related to the Riemann-Roch theorem (see [3]). In the general case Baum and Macpherson [7] have recently given a definition of α_0. In principle this is tantamount to a Riemann-Roch theorem for singular varieties.

§3. Functional Analysis

The K-theory of §2 has a very natural interpretation in terms of operator theory. I recall that a bounded operator F on a complex Hilbert space H is called Fredholm if

$$\dim \operatorname{Ker} F < \infty \text{ and } \dim \operatorname{Coker} F < \infty .$$

Its index is then defined as

$$\text{index } F = \dim \text{Ker } F - \dim \text{Coker } F$$

(note Coker $F \cong$ Ker F^*). If $\mathcal{F}$ denotes the space of all Fredholm operators on H, endowed with the norm topology,

$$\text{index} : \mathcal{F} \to Z$$

gives a bijection on components (i.e. F_1 and F_2 are in the same component of $\mathcal{F}$ if and only if index F_1 = index F_2, and every integer occurs as an index).

If now X is a compact space and $F : X \to \mathcal{F}$ is a continuous map (so that F_x is a continuous family of Fredholm operators) one can define an element index $F \in K^0(X)$. If dim Ker F_x is constant the vector spaces Ker F_x and Coker F_x form two vector bundles over X and their difference in $K^0(X)$ is the index of F. The general case is reduced to this by a deformation. Then one has an <u>isomorphism</u>

$$\text{index} : [X, \mathcal{F}] \to K^0(X)$$

where [,] denotes homotopy classes of maps [1, Appendix].

In [2] I drew attention to the desirability of having a natural analytic definition of $K_0(X)$ and I outlined a possible approach. This consists of looking at triples $\{H_1, H_2, F\}$ where H_i $(i = 1,2)$ are Hilbert spaces which are C(X)-modules (i.e. we have *-homomorphisms $\alpha_i : C(X) \to B(H_i)$, the bounded operators on H_i) and $F : H_1 \to H_2$ is a Fredholm operator with the property that

$$[F,f] = F\,\alpha_1(f) - \alpha_2(f)\,F$$

is compact for every $f \in C(X)$. The group generated by such triples is denoted by Ell (X) (because of the examples furnished by elliptic operators, see §4) and, for X a polyhedron, one can show that there is a natural <u>surjection</u>

$$\text{Ell}(X) \to K_0(X).$$

It is not clear what equivalences one should impose on Ell (X) so as to obtain $K_0(X)$ as quotient. This problem has now been solved by Brown-Douglas-Fillmore in a very satisfactory manner. Somewhat related results have been announced by Kasparov [9]. The BDF theory identifies $K_1(X)$ with a group of extensions of $C(X)$ by the ideal $\mathcal{K}$ of compact operators in $\mathcal{B}(H)$:

$$\text{Ext}\ (C(X),\mathcal{K}) \cong K_1X\ .$$

Equivalently $K_1(X)$ is given by $*$-embeddings of $C(X)$ in the Calkin algebra $\mathcal{A} = \mathcal{B}(H)_{/\mathcal{K}}$ up to unitary equivalence. This can be relativized to the case of a map $\pi : X \to Y$, the corresponding extensions being split over $\pi^* C(Y) \subset C(X)$. Applying this to $X \times S^1 \to X$, where S^1 is the circle, we see that $K_0(X)$ is given by extensions of $C(X \times S^1)$ by $\mathcal{K}$ which are split over $C(X)$. Such an extension, or rather the embedding $C(X \times S^1) \to \mathcal{A}$, gives an **element** of Ell (X), the operator F corresponding to the basic function z on S^1. This explains the relationship of the BDF theory with the approach in [2].

If X is a compact subset of $\mathbb{C}$ the map

$$\text{Ext}\ (C(X),\mathcal{K}) \to K_1(X) = \tilde{K}^0(\mathbb{C} - X) \cong \tilde{H}^0(\mathbb{C}-X;Z)$$

is given explicitly as follows.

Let $C(X) \to \mathcal{A}$ be the embedding corresponding to the given extension and let $T \in \mathcal{B}(H)$ represent (mod $\mathcal{K}$) the image in this embedding of the function $z \in C(X)$ (given by the embedding $X \to \mathbb{C}$). Then X is the essential spectrum of T and so, for $\lambda \in \mathbb{C} - X$, $T - \lambda$ is Fredholm. Hence, by **assigning** to each bounded component U of $\mathbb{C} - X$ the integer index $T - \lambda$, for any $\lambda \in U$, we get an element of $\tilde{H}^0(\mathbb{C}-X,Z)$ as required.

If X is a compact subset of $\mathbb{C}^n = R^{2n}$, there is a natural way to generalize the above which is presumably equivalent to the BDF method. Let $A_1,\dots A_{2n}$ be the 2n operators of $\mathcal{B}$ which represent (mod $\mathcal{K}$) the images of the coordinate functions $x_1,\dots x_{2n}$ under a given embedding $C(X) \to \mathcal{A}$. Let $E_1,\dots E_{2n}$ be the basic Clifford matrices i.e. $2^{n-1} \times 2^{n-1}$ complex matrices such that

$$E_i^2 = -1 \quad E_iE_j = -E_jE_i \quad \text{for} \quad i \neq j.$$

Then the operator

$$F(\mu) = \sum_{i=1}^{2n} (A_i - \mu_i)\, E_i \qquad \mu = (\mu_1,\dots,\mu_{2n})$$

has the property that, modulo $\mathcal{K}$,

$$F(\mu)^2 = - \sum_{i=1}^{2n} (A_i - \mu_i)^2 .$$

Hence, if $\mu \notin X$, $F(\mu)^2$ modulo $\mathcal{K}$ is invertible in $C(X) \subset \mathcal{A}$, and so $F(\mu)$ is Fredholm. Thus

$$\mu \to F(\mu)$$

gives a continuous mapping $R^{2n} - X \to \mathcal{F}$ and so an element of $K^0(R^{2n} - X)$. In fact, for μ large, $F(\mu)$ is actually invertible so that our element lies in $\tilde{K}^0(R^{2n} - X) = K_1(X)$. The BDF theory thus asserts that two extensions are equivalent if and only if the corresponding maps $R^{2n} - X \to \mathcal{F}$ are homotopic.

§4. Differential Equations

Elliptic differential operators on a compact manifold M give rise to Fredholm operators and so tie up globally with the theory discussed in §3. In addition, however, a differential operator D is given by explicit local data - the symbol $\sigma(D)$ - and this naturally defines an element of $K^1(SM)$, where S is the tangent sphere bundle of M. Thus K-theory enters both locally and globally and the index theorem of [4] gives an explicit tie-up between these. In the framework of §3 the situation may be explained as follows. The norm closure $\bar{\mathcal{P}}$ of the algebra of pseudo-differential operators $\mathcal{P}$ of order zero on M gives a natural extension

$$0 \to K \to \bar{\mathcal{P}} \xrightarrow{\sigma} C(SM) \to 0$$

split over C(M) and hence, by the relative BDF theory, anelement of the relative group

$$K_1(SM \to M) = K_0(BM, SM)$$

where BM is the unit tangent ball bundle of M. The index theorem identifies this element in geometrical terms. Modulo torsion we can identify K_0 with even-dimensional homology and the class in

$$H_*(BM, SM) \simeq H_*(M) \simeq H^*(M)$$

is the index class $\mathcal{J}(M)$ of [5]. The full identification requires the index theorem for families of [6]. A heuristic way of identifying the element of $K_0(BM, SM)$ is to pretend that the almost complex structure on BM (directions in M being "real" and those along the fibre being "imaginary") is actually complex algebraic and that we are in a compact situation. Then the fundamental sheaf of BM exists and gives the basic class.

Note that the differential geometry of M, which defines $\mathcal{P}$, is discarded when we go to the norm closure $\bar{\mathcal{P}}$. Thus for more general extensions it is a reasonable programme to try to describe the BDF class of the extension when this has been defined on some dense sub-algebra with more refined structure. The Helton-Howe theory [8] gives a first step in this direction for operators whose bracket is of trace-class. A complete generalization of the index theorem along these lines appears to me to be the main unsolved problem in this area.

It is perhaps worth pointing out one aspect of the index theorem for manifolds. The class in $K_1(SM)$ determined by the extension $\bar{\mathcal{P}}$ is entirely determined by the symplectic structure (or the almost complex structure) of the cotangent bundle TM. This in turn is determined by the Poisson bracket of functions on TM, or alternatively by the bracket of pseudo-differential operators of order one. Thus the first piece of non-commutativity in the algebra in this case determines the extension class. This may be related to the absence of singularities. One might conjecture that reasonable singularities require knowledge of only a "finite part" of the algebra to determine the extension class.

An outstanding problem which a generalization of the index theorem should include is the Hirzebruch Riemann-Roch theorem for a coherent analytic sheaf on a compact complex manifold. The Euler characteristic χ is well-defined and the Chern classes can be defined by the method of [3], so that the Theorem can be formulated. At sent no proof is known except in the projective algebraic case, when resolutions by locally free sheaves exist and reduce the problem to the case of vector bundles.

REFERENCES

1. M.F. Atiyah, K-Theory, Benjamin, 1967.
2. __________, Global theory of elliptic operators, Functional Analysis and related topics, University of Tokyo, 1967.
3. __________ and F. Hirzebruch, The Riemann-Roch theorem for analytic embeddings, Topology 1 (1962), 151-166.
4. __________ and I.M. Singer, The index of elliptic operators I, Ann. of Math. 87 (1968), 484-530.
5. ____________________, The index of elliptic operators III, Ann. of Math. 87 (1968), 546-604.
6. ____________________, The index of elliptic operators IV, Ann. of Math. 93 (1971), 119-138.
7. P. Baum and Macpherson (to appear).
8. J.W. Helton and R. Howe, Integral operators: commutators, traces, index and homology, Springer Lecture Notes No. 345, p. 141-209.
9. G.G. Kasparov, Generalized index of elliptic operators, Functional Analysis 7 (1973), 82-83.

CHARACTERIZING EXT(X)

Lawrence G. Brown, Purdue University

I will speak on four specific topics:

1. The meaning of the statement, "Ext(X) is (reduced) Steenrod K-homology".

2. The universal coefficient theorem.

3. Strong homotopy invariance.

4. Open questions.

§1. Steenrod K-homology.

Among the homology and cohomology theories in general use, the one that seems best adapted to not necessarily smooth compact spaces is the Céch theory. This has two useful properties that the singular theory does not have: It is continuous; that is, it is compatible with inverse limits; and it satisfies a stronger form of the excision axiom. For this reason Čech cohomology is clearly the preferred theory for compact spaces, but there is a problem with Čech homology. The trouble is that since inverse limits of exact sequences need not be exact, the "long exact sequence" in Čech homology theory is not in general exact. There is another homology theory for compact metric spaces which eliminates this defect for a reasonable price. It was originated by Steenrod in [10] and axiomatized by Milnor [8], who named it Steenrod homology. The axioms of Steenrod homology consist of the Eilenberg-Steenrod axioms together with two additions:

1. The strong excision axiom (just as in Céch theory).

2. The cluster set axiom (which is a special case of the continuity axiom of Čech theory).

The cluster set axiom states that if X is the union of closed subsets X_n, whose diameters tend to 0 and which have one point in common and are otherwise disjoint, then the homology of X is the product of the homology of the X_n's. Now Ext(X) satisfies all the axioms of Steenrod homology, except the dimension axiom, of course, and hence is a generalized Steenrod homology theory.

The statement, "Ext(X) is (reduced) Steenrod K-homology" means that it is the Steenrod homology theory which "goes with" K-theory (which is a generalized cohomology theory), but at this point I have to confess that I do not know the precise meaning of the statement, though I am certain it is true. There is a precise theory, known to topologists, which identifies the generalized homology theory that goes with any cohomology theory on finite complexes, and Ext(X) for X a finite complex is reduced K-homology. But unfortunately there is no known uniqueness theorem for extensions of a homology theory on finite complexes to a Steenrod theory on compact metric spaces. There are certain pairings, such as the cap product, which usually exist between corresponding homology and cohomology theories. One might hope to use these to explain the statement I am discussing, but so far as I know there is no uniqueness theorem based on these pairings in the category of compact metric spaces. However, these pairings between K-theory and Ext(X) do exist; and I believe that, if anything, the relations between K-theory and Ext(X) are richer than one would expect from axiomatic considerations.

For example, the cap product gives a module action of $K^0(X)$ on Ext(X), and the relative version gives an action of $K^0(X)$ on Ext(X/A). More concretely, any vector bundle on X, even if it does not come from a bundle on X/A, gives an endomorphism of Ext(X/A). What I find surprising is that one need not even have a bundle on X -- it is sufficient to have a bundle on the locally compact space X - A. In particular, if X - A is contractible, the cap product pairing between $\tilde{K}(X)$ and Ext(X/A) is trivial. Now in simple cases, such as the case where X is the suspension of a space and A is the north pole, this last fact follows from axiomatic considerations; but so far as I know (I disclaim any topological expertise), there is no axiomatic proof of this for general compact metric spaces.

§2. The universal coefficient theorem.

The universal coefficient theorem asserts the following natural exact sequence:

$$(1) \quad 0 \to \mathrm{Ext}^1_Z(K^{q+1}(X), Z)) \to \mathrm{Ext}_q(X) \xrightarrow{\gamma_\infty} \mathrm{Hom}(K^q(X), Z)) \to 0.$$

This sequence splits non-canonically, and hence if we know the K-theory of X, it yields Ext(X) up to non-canonical isomorphism. The γ_∞ appearing here is the one described in Douglas' talk, the basic "index invariant".

To prove (1), one uses Milnor's $\lim^{(1)}$ sequence, which states that if X is the inverse limit of $\{X_n\}$, then there is an exact sequence:

$$(2) \quad 0 \to \lim^{(1)}(\mathrm{Ext}_{q-1}(X_n)) \to \mathrm{Ext}_q(X) \to \underleftarrow{\lim}(\mathrm{Ext}_q(X_n)) \to 0.$$

Here $\lim^{(1)}$ is the first derived functor of inverse limit -- roughly speaking it measures the extent to which inverse limits of exact sequences fail to be exact. Since the continuity axiom is precisely this sequence with 0 instead of $\lim^{(1)}$, $\lim^{(1)}$ is the "reasonable price" to which I referred before.

Now $\mathrm{Ext}_q(X)$ is originally defined only for $q \leq 1$ ($q = 1$ gives the basic group, Ext(X)). Thus Milnor's axiomatic proof initially yields (2) only for $q \leq 0$. This enables one to prove the final step of Bott periodicity, the surjectivity of Per: $\mathrm{Ext}_{q-2}(X) \longrightarrow \mathrm{Ext}_q(X)$, for $q \leq 0$. Bott periodicity can then be extended to $q = 1$, making it possible to define Ext_q for $q > 1$, and yielding (2) in complete generality.

Now if each X_n is a simplicial complex, which is always possible, the $\lim^{(1)}$ group is divisible and hence (2) splits. Since $\mathrm{Ext}(X_n)$, for X_n a simplicial complex, is "known" on topological grounds, (2) gives another method for characterizing Ext(X).

To get (1), one uses a map $\varkappa$:Ker $\gamma_\infty \to \mathrm{Ext}^1_Z(K^0(X),Z)$. (From now on, I assume q = 1 for simplicity.) $\varkappa$ is defined by means of algebraic K-theory, and I will say something later about algebraic K-theory and the definition of $\varkappa$. For now all we need to know is that $\varkappa$ exists and satisfies the naturality properties necessary for our purposes. Using $\varkappa$ one can prove (1) for X a simplicial complex, by induction on the number of simplices. More precisely, one proves that γ_∞ is surjective and $\varkappa$ is an isomorphism. It is apparently not possible to prove the surjectivity of γ_∞ by induction on the number of simplices without $\varkappa$, although this surjectivity was previously known ([11]). (2) is used to prove that γ_∞ is surjective and $\varkappa$ is an isomorphism for general X.

There is another exact sequence:

$$(3) \qquad 0 \to \mathrm{Ext}^1_Z(K^*(X),\ \mathrm{Ext}_*(Y)) \to \mathrm{Ext}_*(X \wedge Y) \to \mathrm{Hom}(\tilde{K}^*(X),\ \mathrm{Ext}_*(Y)) \to 0.$$

I don't know if this splits (because I don't know whether the Künneth sequence in K-theory splits*), but modulo splitting (3) tells us how to compute Ext(X×Y) and is thus a substitute for a Künneth theorem. I don't know whether there is an actual Künneth theorem (which would involve tensor products). I think not, but I can't prove it. The proof of (3) makes use of the proof of the Künneth theorem of K-theory given by Atiyah [3].

§3. Algebraic K-theory.

Algebraic K-theory gives a sequence of covariant functors, K_i i = 0, 1,2 ... , from rings with identity to abelian groups. Also, for I an ideal in R, there are "relative groups $K_i(I)$ and a long exact sequence:

$$\begin{array}{c} \vdots \\ K_1(I) \longrightarrow K_1(R) \longrightarrow K_1(R/I) \\ \swarrow \\ K_0(I) \longrightarrow K_0(R) \longrightarrow K_0(R/I) \end{array}$$

$K_i(I)$ depends on the embedding of I in R except for i = 0.

*As of April 28, 1975 no one has resolved this question for me except for finite complexes.

The definition of K_i is purely algebraic, of course; and at least for low values of i (which is all that I know anything about) the kind of algebra used is natural for operator theory. It involves idempotents, similarity and such; and in case R is a C^*-algebra, the idempotents can be assumed to be self-adjoint and similarity can be replaced by unitary equivalence. It also involves $M_n(R)$, the ring of $n \times n$ matrices with entries in R.

Now $K_o(R)$ is usually defined by means of projective R-modules, but the definition can be interpreted concretely. Elements of $K_o(R)$ can be defined as formal differences of idempotents in $M_n(R)$ (where n can vary) modulo an equivalence relation. The main operation that leads to equivalent idempotents is a similarity induced by an invertible element of $M_n(R)$. If I is an ideal in R, $K_o(I)$ can be defined similarly. One can use formal differences of idempotents in $M_n(R)$ which are the same modulo I, and the only similarities allowed are those induced by matrices congruent to the identity modulo I. I am oversimplifying a little bit.)

At this point I wish to discuss the fact that $K_o(\mathcal{K}) \cong Z$, where $\mathcal{K}$ is the ideal of compact operators on Hilbert space. Using the description of $K_o(I)$ just given, one gets this directly from the theory of essential codimension introduced in [6], section 4.9. I personally find it convenient to use this way of getting the isomorphism, but it also follows directly from the long exact sequence:

$$\cdots \to K_1(\mathcal{L}(\mathcal{H})) \xrightarrow{j} K_1(\mathcal{A}) \to K_o(\mathcal{K}) \to K_o(\mathcal{L}(\mathcal{H})) = 0.$$

Here $\mathcal{A}$ is the Calkin algebra, and the basic theory of the Fredholm index allows us to identify the cokernel of j with Z.

$K_1(R)$ can be defined as the group of invertibles in $M_n(R)$ (where n can vary) modulo the commutator subgroup. Since commutator theory, including multiplicative commutator theory, is already an established part of operator theory, I won't say anything more about this.

One more fact that we need is the relation between algebraic and topological K-theory, which is seen by taking $R = C(X)$. There are natural maps $K_i(C(X)) \to K^i(X)$ (at least for $i = 0,1,2$). For $i = 0$, the map is an isomorphism; for $i = 1$, it is surjective and its kernel is "under control". (For $i = 2$, the kernel is no longer under control -- there is no Bott periodicity for algebraic K-theory.)

Now to define χ, let $0 \to \mathcal{K} \to \mathcal{U} \to C(X) \to 0$, be an extension in the kernel of γ_∞. Applying the algebraic K-theory long exact sequence, we get:

$$\cdots \to K_1(C(X)) \xrightarrow{\partial} K_0(\mathcal{K}) \to K_0(\mathcal{U}) \xrightarrow{P} K_0(C(X)) \to 0$$

$$K_0(\mathcal{K}) = Z, \qquad K_0(C(X)) = K^0(X)$$

The surjectivity of P comes from a well-known property of the Calkin algebra -- namely that every projection in $\mathcal{a}$ is the image of a projection in $\mathcal{L}(H)$. Using the identifications I have described, we see that ∂ is just γ_∞. Thus we have $0 \to Z \to K_0(\mathcal{U}) \to K^0(X) \to 0$, and $K_0(\mathcal{U})$ "is" an element of $\mathrm{Ext}^1_Z(K^0(X),Z)$.

§4. Strong homotopy invariance.

Let τ_t, $0 \leq t \leq 1$, be a family of $*$-isomorphisms of $C(X) \to \mathcal{a}$, continuous in the sense that $\tau_t(f)$ is continuous in t for each $f \in C(X)$. I wish to show that each τ_t defines the same element, $[\tau_t]$, of Ext(X). In case all the images $\tau_t(C(X))$ lie in a commutative subalgebra of $\mathcal{a}$, we recover the usual homotopy invariance principle.

Now it is easy to see that $\gamma_\infty([\tau_t])$ does not depend on t. Therefore we may assume that each $[\tau_t]$ is in the kernel of γ_∞, and the problem is to prove that $\chi([\tau_t])$ is independent of t.

$\chi([\tau_t])$ is a group extension of Z by the (countable) group $K^0(X)=$

$K_o(C(X))$ and is described by a 2-cocycle $\rho_t : K_o(C(X)) \times K_o(C(X)) \to Z$. To compute ρ_t, let α, $\beta \in K_o(C(X))$ be defined by projections p', p" $\in M_n(C(X))$ $\tau_t(p')$ and $\tau_t(p'')$ are projections in $\mathcal{Q}$, and we have to lift them to projections P_t', P_t'' in $\mathcal{L}(H)$. P_t' and P_t'' can be chosen to depend continuously on t. (Here I am using the fact that $M_n(\mathcal{L}(\mathcal{H})) \simeq \mathcal{L}(\mathcal{H})$ and $M_n(\mathcal{Q}) \cong \mathcal{Q}$; and I am suppressing "n" in my notation.) Similarly, let p define $\alpha + \beta$, and P_t lift $\tau_t(p)$. To find $\rho_t(\alpha,\beta)$ we have to compare P_t with $\begin{pmatrix} P_t' & 0 \\ 0 & P_t'' \end{pmatrix}$. Now since p and

$\begin{pmatrix} p' & 0 \\ 0 & p'' \end{pmatrix}$ define the same element of $K_o(C(X))$, there are certain concrete operations that transform one into the other. If we apply the same operations to P_t and $\begin{pmatrix} P_t' & 0 \\ 0 & P_t'' \end{pmatrix}$ (in such a way as to preserve the continuity in t) we arrive at two projections which are the same modulo $\mathcal{K}$. The essential codimension of these is $\rho_t(\alpha,\beta)$. Since essential codimension is continuous, ρ_t does not depend on t.

I do not know whether the covering homotopy property (pp. 121-122 of [6]) is true. But Conway [7] has proved it for some special cases and these cases were used above in performing the liftings from $\mathcal{Q}$ to $\mathcal{L}(\mathcal{H})$ so as to preserve continuity in t.

§5. Generalizations and open questions.

All of the C^*-algebras, extensions of $\mathcal{K}$ by C(X), which arise in defining Ext(X) have the same dual -- namely the result of adjoining a "generic point" to X. Any algebra with this dual is an extension of $\mathcal{K}$ by $\mathcal{C}$, where $\mathcal{C}$ is some C^*-algebra with dual X. It is natural to try to classify these extensions. For some special cases (other than the original case $\mathcal{C} = C(X)$) these extensions, under weak equivalence, are known to yield the same group, Ext(X). There are some other cases where they yield groups, equivalent to Ext(X) modulo torsion, which presumably correspond to the twisted K-groups.

Now one of the techniques used in the theory of Ext(X) is to perform topological manipulations on the compact metric space X. Unfortunately, one doesn't have the same freedom to use this technique when C(X) is replaced with $\mathcal{C}$. Therefore more abstract C^*-algebraic techniques would be desirable. I want to raise four specific problems of this sort:

1. Is Ext($\mathcal{C}$) a group?
2. Does Ext($\mathcal{C}$) have strong homotopy invariance?
3. Can every $\tau:\mathcal{C}_1 \otimes \mathcal{C}_2 \to \mathcal{A}$ be extended to $\tau':\mathcal{C}_1' \otimes \mathcal{C}_2' \to \mathcal{A}$?
4. Can every $\tau:\mathcal{J} \to \mathcal{A}$ be extended to $\tilde{\tau}:\mathcal{C} \to \mathcal{A}$?

Here $\mathcal{C}$, $\mathcal{C}_1$ and $\mathcal{C}_2$ are C^*-algebras, not necessarily as above (any reasonable hypotheses on $\mathcal{C}$ would be all right, so long as they are stated abstractly), $\mathcal{C}_1'$ and $\mathcal{C}_2'$ are the C^*-algebras obtained by adjoining an identity to $\mathcal{C}_1$ and $\mathcal{C}_2$, and $\mathcal{J}$ is an ideal in $\mathcal{C}$ such that $\mathcal{C}/\mathcal{J}$ is either $\mathcal{K}$ or $M_n(\mathbb{C})$.

Douglas has already mentioned the new proof that Ext(X) is a group, due to Arveson [1] and Davie. This illustrates the kind of answer I want to these questions, though so far as I know it does not answer 1. for $\mathcal{C}$ more general than C(X).

The importance of 3. and 4. is not as obvious as that of 1. and 2., and I do not guarantee that solution of 3., in particular, will prove useful.

Both 3. and 4. are special cases of the problem of extending $\tau:\mathcal{C} \to \mathcal{A}$ to $\tilde{\tau}:\tilde{\mathcal{C}} \to \mathcal{A}$, where $\mathcal{C}$ is a C^*-algebra without identity and $\tilde{\mathcal{C}}$ is between $\mathcal{C}$ and its double centralizer algebra. Other special cases might also be of interest, but the general case is definitely false (as can be seen already from the case where $\mathcal{C}$ is commutative).

References

1. W.B. Arveson, A note on essentially normal operators, Aarhus University, preprint, 1974.

2. M.F. Atiyah, Global theory of elliptic operators. Internat. Conf.

on Functional Analysis and Related Topics (Tokyo, 1969), Univ. of Tokyo Press, Tokyo 1970, pp. 21-30.

3. __________, K-Theory, Benjamin, New York, 1967.

4. L.G. Brown, Operator algebras and algebraic K-theory, Bull. Amer. Math. Soc. (to appear).

5. __________, R.G. Douglas, and P.A. Fillmore, Extensions of C^*-algebras, operators with compact self-commutators, and K-homology, Bull. Amer. Math. Soc. 79 (1973), 973-978.

6. __________, Unitary equivalence modulo the compact operators and extensions of C^*-algebras, Proc. Conf. on Operator Theory, Lecture Notes in Math. vol. 345, Springer-Verlag, New York, 1973, pp. 58-128.

7. J. Conway, On the Calkin algebra and the covering homotopy property, preprint.

8. J. Milnor, On the Steenrod homology theory, mimeographed notes, Berkeley, 1961.

9. __________, Introduction to Algebraic K-Theory, Ann. of Math. 72 (1971).

10. N. Steenrod, Regular cycles on compact metric spaces, Ann. of Math. 41 (1940), 833-851.

11. D. Sullivan, Geometric topology seminar notes, Princeton, 1967.

ALMOST COMMUTING ALGEBRAS

Richard W. Carey, University of Kentucky and Joel D. Pincus, State University of New York at Stony Brook

We shall give a brief account of certain results and problems that arise in connection with algebras of operators generated by elements that "almost commute".

Let $\mathcal{M}$ be a weakly closed self-adjoint subalgebra of $\mathcal{L}(\mathcal{H})$, equipped with a normal semi-finite trace τ. We will restrict our attention in this note to the cases where $\mathcal{M}$ is either $\mathcal{L}(\mathcal{H})$ itself or to the case where $\mathcal{M}$ is a type II_∞ factor. In either case we will denote the trace ideal by $\mathcal{J}_\tau$.

Let $T \in \mathcal{M}$ and assume that $T^*T-TT^* = 2C \in \mathcal{J}_\tau$. Our method revolves around the study of properties of a certain operator valued function, the determining function of T, and we will survey some of the facts about this function and particularly the properties of an associated object, the so-called principal function.

The principal function is an extension of the Fredholm index of $T-(\mu+i\nu)I$ to the entire plane in case $\mathcal{M}$ is $\mathcal{L}(\mathcal{H})$ and it is an extension of the Breuer relative index of $T-(\mu+i\nu)$, [1], in the type II_∞ case. In both cases the assignment of principal function to T is invariant under unitary transformations in $\mathcal{M}$ and under perturbations of T by elements in $\mathcal{J}_\tau$.

We will begin with a survey of known facts, and will end with an account of some results which are new.

1. Traces and Holomorphic Functions.

Suppose that T is completely non-normal and $T = U+iV$ where U,V are self-adjoint and belong to $\mathcal{M}$ and $VU-UV = -iC \in \mathcal{J}_\tau$.

Factor C in the form $\hat{C}^2 = \quad = \hat{C}^*J\hat{}$ where J is self-adjoint and unitary, and set $J = P_+-P_-$ where $P_\pm$ are projections onto the spectral spaces of J corresponding to the spectral points 1 and -1.

The present theory is build on a simple idea: The operator

$$(1.1)\qquad E(\ell,z) = 1 + \frac{1}{i}\hat{C}(V-\ell)(U-z)^{-1}\hat{C}$$

defined for Im $\ell \neq 0$, Im $z \neq 0$, "undoes" the non-commutativity of V and U in the sense that

$$(1.2)\qquad (V-\ell)^{-1}(U-z)^{-1}\hat{C} = (U-z)^{-1}(V-\ell)^{-1}\hat{C}\, E(\ell,z).$$

Furthermore, $E(\ell,z)$ is a complete unitary invariant for T.

It is natural to consider the C* algebra $\mathcal{A}$ generated by T, and to attempt to find out what information about $\mathcal{A}$ is contained in properties of $E(\ell,z)$, the determining function of T.

When $\mathcal{M}$ is $\mathcal{L}(\mathcal{H})$ and τ is the usual trace, it is possible to form the determinant of $E(\ell,z)$.

Theorem 1.1. [10], [11], [2]. There exists a summable function $g(\nu,\mu):R^2 \to R$ such that

$$\det E(\ell,z) = \exp \frac{1}{2\pi i} \iint g(\nu,\mu) \frac{d\nu}{\nu-\ell} \frac{d\mu}{\mu-z}.$$

This theorem is not applicable to the type II_∞ case because det $E(\ell,z)$ is not then defined. However, it is not difficult to show that the following equation is equivalent to the equality of the theorem:

$$(1.3)\qquad \tau[(U-z)^{-1}(V-\ell)^{-1}(U-z)^{-1}[V,U]] = \frac{i}{2\pi} \iint g(\nu,\mu) \frac{d\nu}{\nu-\ell} \frac{d\mu}{(\mu-z)^2}$$

In this form the essential idea underlying theorem 1.1 makes sense in a type II_∞ factor.

Theorem 1.2. [11],[3]. There exists a unique summable function $g(\nu,\mu)$ the principal function of the pair $\{V,U\}$, such that

$$\tau[(U-z)^{-1}(V-\ell)^{-1}(U-z)^{-1}[V,U]]$$

$$= \frac{i}{2\pi} \iint g(\nu,\mu) \frac{d\nu}{\nu-\ell} \frac{d\mu}{(\mu-z)^2}$$

The existence statement made here is far from trivial in the

II_∞ case, and even in the $\mathcal{L}(\mathcal{H})$ case it is absolutely crucial for us to deal with a function and not simply some linear functional or measure. In this connection, see Helton and Howe [7] where a study is made to abstract certain properties of the present theory.

We will sketch later the reasons for the existence of the principal function.

2. Symbols and $g(\nu,\mu)$: The Type I_∞ Case.

Now we turn to the question of the computability of $g(\nu,\mu)$. Assume until further stated that we are in the $\mathcal{L}(\mathcal{H})$ case.

Let $\mathfrak{M}(U)$ be the C*-algebra generated by operators having a trace class commutator with U.

Theorem 2.1. If $A \in \mathfrak{M}(U)$, then the strong limits

$$\text{s-}\lim_{t\to\pm\infty} e^{itU} A\, e^{-itU} P_a(U) = S_\pm(U;A) \text{ exist.}$$

We call these operators, which commute with U, the symbols of A with respect to U.

We list some properties of the symbols:

If A,B are in $\mathfrak{M}(U)$, then

(2.1) $\quad S_\pm(U;A^*) = S_\pm(U;A)^*$

(2.2) $\quad S_\pm(U;A+B) = S_\pm(U;A) + S_\pm(U;B)$

(2.3) $\quad S_\pm(U;AB) = S_\pm(U;A)S_\pm(U;B)$

(2.4) $\quad \|S_\pm(U;A)\| \le \|A\|$

(2.5) $\quad$ kernel $S_\pm(U;\cdot)$ contains the compact operators.

Since $S_\pm(U;A)$ is in the commutant of U it is decomposable in a direct integral space $\int \oplus \mathcal{H}_\lambda d\lambda$ which diagonalizes the absolutely continuous part of U.

Thus

$$S_\pm(U;A) = \int \oplus\, S_\pm(U;A)(\lambda)d\lambda$$

where $S_{\pm}(U;A)(\lambda)$ operate on H_λ.

These operators can be quite explicitly computed in many cases, and their unitary invariants remain unaltered if a trace class operator K is added to U.

The following result is a generalization of the main theorem in [10] to the case where C can have eigenvalues which are both positive and negative.

<u>Theorem 2.2.</u> Let E_λ and F_λ be the spectral resolutions of U and V respectively. Then

$$E^{-1}(\ell,\lambda-io)E(\ell,\lambda+io) = 1 + 2\frac{d}{d\lambda}[\hat{C}E_\lambda(S_+(U;V)-\ell)^{-1}\hat{C}]$$

$$E^{-1}(\ell,\lambda+io)E(\ell,\lambda-io) = 1 - 2\frac{d}{d\lambda}[\hat{C}E_\lambda(S_-(U;V)-\ell)^{-1}\hat{C}]$$

$$E(\nu+io,z)E^{-1}(\nu-io,z) = 1 + 2\frac{d}{d\nu}[\hat{C}F_\nu(S_+(V;U)-z)^{-1}\hat{C}]$$

$$E(\nu-io,z)E^{-1}(\nu+io,z) = 1 - 2\frac{d}{d\nu}[\hat{C}F_\nu(S_-(V;U)-z)^{-1}\hat{C}]$$

The derivatives are taken in the trace norm topology.

Furthermore

$$\det[1 + 2\frac{d}{d\lambda}\hat{C}E_\lambda S_+(U;V-\ell)^{-1}\hat{C}]$$

$$= e^{\int g(\nu,\lambda)\frac{d\nu}{\nu-\ell}}$$

and

$$\det[1 + 2\frac{d}{d\nu}\hat{C}F_\nu S_+(V;U-z)^{-1}\hat{C}]$$

$$= e^{\int g(\nu,\lambda)\frac{d\lambda}{\lambda-z}}.$$

This relation implies that $g(\nu,\mu)$ is dominated by multiplicity.

<u>Proposition 2.3.</u> $|g(\nu,\mu)| \leq \min\{m(U,\mu),m(V,\nu),\mathrm{Rank}[V,U]\}$ where $m(U,\mu)$ and $m(V,\nu)$ are the multiplicity functions for the operators U_a and V_a, respectively.

The simplest example is obtained by taking $Ux(\lambda) = \lambda x(\lambda)$ on $L^2(a,b)$ and $Vx(\lambda) = A(\lambda)x(\lambda) + \frac{1}{\pi i}\int_a^b \frac{\bar{k}(\lambda)k(\mu)}{\mu-\lambda}x(\lambda)d\mu$ for $A(\lambda)$ real in

$L^\infty(a,b)$ and $k(\lambda)$ in $L^\infty(a,b)$.

In this case $S_{\pm}(U,V)(\lambda) = A(\lambda) \mp |k(\lambda)|^2$ and $g(\nu,\lambda)$ is the characteristic function of the set of points (ν,λ) such that $A(\lambda) - |k(\lambda)|^2 < \nu < A(\lambda) + |k(\lambda)|^2$.

The principal function can be explicitly computed from the symbols because, for $\operatorname{Im} \ell \neq 0$

$$\det[1 + 2\frac{d}{d_\lambda}(\hat{c}\, E_\lambda (S_+(U;V) - \ell)^{-1}\hat{c})]$$

$$= \det(1 + (S_-(U;V)(\lambda) - S_+(U;V)(\lambda))S_+(U;V-\ell)^{-1}(\lambda))$$

$$= \exp \int g(v,\lambda)\frac{d\nu}{\nu - \ell}$$

See [5].

The fact that $g(\nu,\lambda)$ is the index of $U + iV - (\mu + i\lambda)1$ was conjectured generally by the second author in 1970, but only established originally under heavy restrictions [15]. The general proof was given in [11], [4], the extension to von Neumann algebras was given in [3]. A related result is in [7].

We have shown how $g(\nu,\mu)$ is defined and computed, we turn now to a discussion of its utility. We now drop the requirement that $\mathcal{M} = \mathcal{L}(\mathcal{H})$.

3. The Cartesian Functional Calculus.

Let $M(R^2)$ be the space of complex measures ω on R^2 with $\iint(1 + |t|)(1 + |s|)d|\omega(t,s)| < \infty$. Let $\hat{M}(R^2)$ be the set of all functions of the form $\iint e^{itx+isy}d\omega(s,t)$. With $F(x,y) \in \hat{M}(R^2)$. We associate the operator $F(U,V) = \iint F(x,y)dE_x df_y$.

This association of operators with functions defines what we call a Mikhlin-Weyl functional calculus.

Theorem 3.1. If $F_1(x,y)$ and $F_2(x,y) \in \hat{M}(R^2)$, then $F_1(U,V)$, $F_2(U,V) \in A$ and $[F_1(U,V), F_2(U,V)]$, $F_1(U,V)F_2(U,V) - (F_1F_2)(U,V)$, $F_1(U,V)^* - \bar{F}_1(U,V)$ are all in $\mathcal{I}_\tau$.

We note that if $F \in \hat{M}(R^2)$, then $F(x,y)$ has continuous partial derivatives which are bounded on R^2. Also, if F is a function on R^2 having compact support with partial derivatives satisfying a Holder condition with exponent greater than 1/2, then $F \in \hat{M}(R^2)$.

Suppose that $\alpha(x,y)$ and $\beta(x,y)$ are real valued functions in $\hat{M}(R^2)$. The corresponding operators $\alpha(U,V)$, $\beta(U,V)$ have a commutator in $\mathcal{J}_\tau$, and up to a term in $\mathcal{J}_\tau$, are self adjoint.

Let Ω be the map $(\nu,\mu) \to (\beta,\alpha)$. It is a nice fact that the principal function associated, by the previous theorem, to the pair of operators $\alpha(U,V)$, $\beta(U,V)$ -- which we will denote by g_Ω -- is easily computed.

Theorem 3.2.

$g_\Omega(\beta,\alpha) = \sum_{(\nu_1,\mu_1)} [\mathrm{Sgn}(\text{Jacobian } \Omega)(\nu_i,\mu_i)] g(\nu_i,\mu_i)$ the summation being extended over those points (ν_i,μ_i) for which $\Omega(\nu_i,\mu_i) = (\beta,\alpha)$.

Since $g_\Omega(y,x)$ is the index of $\alpha(U,V) + i\beta(U,V) - (x+iy)I$, Theorem 3.2 provides a rather explicit index formula.

This result is a simple consequence of the following "Poisson-Bracket" formula, a consequence of equation (1.3) which follows by taking residues at infinity. [11] see also [3].

Theorem 3.3. If F and $H \in \hat{M}(R^2)$, then

$$\tau[F(U,V),H(U,V)] = \frac{i}{2\pi} \iint \frac{\partial(F,H)}{\partial(\mu,\nu)} g(\nu,\mu)\, d\nu\, d\mu .$$

The second author also discovered in [11] the simple identity

$$\det[e^A \; e^B \; e^{-A} \; e^{-B}] = e^{\operatorname{trace}[A,B]}$$

This gives an alternate form of theorem 3.3 if we take $A = F_1(U,V)$, $B = F_2(U,V)$ in $\hat{M}(R^2)$.

4. Intertwining Partial Isometries.

Theorem 3.2 does not, however, exhaust the transformation

properties associated with the principal function.

There are important operators naturally associated to $T \in A$ which are not obtained via the Cartesian functional calculus. For example, let $T = W(T^*T)^{1/2}$ where W is a partial isometry with initial space equal to the range of T^* and final space equal to the range of T.

If $0 \notin$ the essential spectrum of T, then we can use Theorem 3.1 to show that

$$\iint \frac{x + iy}{\sqrt{x^2 + y^2}} \, dE_x dF_y$$

will differ from W by a trace class operator. But this may not be the case if 0 belongs to the essential spectrum of T.

In order to treat operators like W, a new layer has to be added to the theory, and we start all over again.

Suppose that W is a partial isometry for which $WV = (V+D)W$ where $V = V^*$, $D = D^*$ and W are in $\mathfrak{M}$ and $D \in \mathcal{J}_\tau$. Let the initial and final spaces of W be called $H_i(W)$ and $H_f(W)$ respectively. Then $H_i(W)$ is invariant for $V + D$.

Define the Hilbert space

$$\tilde{H} = \ldots \oplus H_f(W)^\perp \oplus H_f(W)^\perp \oplus H \oplus H_i(W)^\perp \oplus H_i(W)^\perp \oplus \ldots$$

and define $\tilde{W}$ so that $\tilde{W}\tilde{X} = \tilde{W} < \ldots, X_{-2}, X_{-1}, X_0, X_1, X_2, \ldots > =$ $< \ldots, X_{-3}, X_{-2}, X_{-1} + WX_0, P_R X_0, X_1, \ldots >$ where P_r is the projection of H onto $H_i(W)^\perp$.

Further, let

$$\tilde{V}\tilde{X} = < \ldots, (V + D)X_{-2}, (V + D)X_{-1} VX_0, VX_1, VX_2, \ldots >$$

$$\tilde{D}\tilde{X} = < \ldots, 0, 0, DX_0, 0, 0, \ldots > .$$

Then we can see that $\tilde{W}\tilde{V} = (\tilde{V} + \tilde{D})\tilde{W}$ and $\tilde{W}$ is the minimal unitary dilation of W.

Now we define the so-called cylinder functional calculus.

Let $M(Z \times R)$ be the collection of Borel measures μ on $Z \times R$

for which $\int_{Z\times R}(1+|n|)(1+|t|)d|\mu(n,t)| < \infty$. Let $\hat{M}(S \times R)$ be the set of characteristic functions of measures in $M(Z \times R)$ of the form

$$F(e^{i\theta},x) = \int_{Z\times R} e^{in\theta} e^{itx} d\mu(n,t) ,$$

and define the operator

$$F(\tilde{W},\tilde{V}) = \int_{Z\times R} \tilde{W}^n t^{itV} d\mu(n,t)$$

in $\mathcal{L}(\tilde{\mathcal{H}})$.

Theorem 4.1. The map

$$F \to PF(\tilde{W},\tilde{V})|_{\mathcal{H}} ,$$

where P is the projection of $\tilde{\mathcal{H}}$ onto $\mathcal{H}$ defines a complex * homomorphism of $\hat{M}(S^1 \times R)$ onto $\mathcal{L}(\mathcal{H})$ modulo $\mathcal{I}_\tau$. Thus, if F and $H \in \hat{M}(S^1 \times R)$, we have

1) $PF(\tilde{W},\tilde{V})^*|_{\mathcal{H}} - P\bar{F}(\tilde{W},\tilde{V})|_{\mathcal{H}} \in \mathcal{I}_\tau$

2) $PF(\tilde{W},\tilde{V})|_{\mathcal{H}} \cdot PH(\tilde{W},\tilde{V})|_{\mathcal{H}} - P(FH)(\tilde{W},\tilde{V})|_{\mathcal{H}} \in \mathcal{I}_\tau$

3) $[PF(W,V)|_{\mathcal{H}}, PH(\tilde{W},\tilde{V})|_{\mathcal{H}}] \in \mathcal{I}_\tau$

In analogy with Theorem 1.2, we have

Theorem 4.2. There exists a unique real valued function $g_{(W,V)}(\lambda,\omega)$ defined on the cylinder $(-\infty,\infty) \times \{\omega: |\omega| = 1\}$ such that

$$\frac{1}{2\pi i}\iint g_{(W,V)}(\lambda,\omega)\frac{1}{\lambda-\ell}\frac{1}{(\omega-z)^2}\,d\lambda d\omega$$
$$= \tau[\hat{D}\tilde{W}(\tilde{W}-z)^{-1}(\tilde{V}-\ell)^{-1}(\tilde{W}-z)^{-1}\hat{D}].$$

Here $\hat{D}$ has been defined so that $\hat{D} \in \mathcal{M}$, $\hat{D}^2 = \tilde{D}$ and $\hat{D}\hat{D}^* = \hat{D}^*\hat{D} = |\hat{D}|$.

The proof of this theorem is very much like the proof of theorem 1.2, but instead makes use of the function

$$\omega(\ell,z) = 1 + \hat{D}\tilde{W}(\tilde{W}-z)^{-1}(\tilde{V}-\ell)^{-1}\hat{D} .$$

See [3], [4] and especially [6] for further information about these functions, and the appropriate version of Theorem 3.3.

Much of the interest in the cylinder principal function $g_{(W,V)}(\lambda,\tau)$ comes from the fact that

$$\delta(\lambda) = \frac{1}{2\pi}\int_0^{2\pi} g_{(W,V)}(\lambda,e^{i\theta})\,d\theta \tag{4.1}$$

where $\delta(\lambda)$ is the so-called spectral displacement function or phase shift corresponding to the scattering problem $V \to V + D$.

Thus, we can prove that

$$\tau\{(V-\ell)^{-1} - (V + D - \ell)^{-1}\} = \int \frac{\delta(\lambda)}{(\lambda-\ell)^2}\,d\lambda \tag{4.2}$$

and thus that

$$\tau\{F(V) - F(V + D)\} = \int F'(\lambda)\delta(\lambda)\,d\lambda \tag{4.3}$$

if $F(\lambda)$ is differentiable.

In case $\mathcal{M}$ is a type II_∞ factor, the existence of a spectral displacement function $\delta(\lambda)$ so that this last equation holds, is a new result of interest in its own right.

<u>Proposition 4.3.</u> $|g(\lambda,e^{i\theta})| \le \min\{m(V,\lambda) + |\delta(\lambda)|, m(W,e^{i\theta}), \mathrm{Rank}[W,V]\}$ where $m(V,\cdot)$ and $m(W,\cdot)$ are the multiplicity functions for the operators V_{ac} and W_{ac} respectively.

If $F \in \hat{M}(Z \times R)$ is unimodular and $H \in \hat{M}(Z \times R)$ is real valued, then there is a corresponding cylinder principal function $g_{(F,H)}(\lambda,\omega)$.

Thus we also obtain an analog of Theorem 3.2 expressing $g_{(F,H)}$ in terms of $g_{(W,V)}$.

When $\mathcal{M}$ is $\mathcal{L}(\mathcal{H})$ the principal function $g_{(W,V)}$ can be obtained in terms of symbols. We have

$$\det[S_-(V;W - z)(\lambda)S_+(V;W - z)^{-1}(\lambda)] = \exp \int_{|\tau|=1} g(\lambda,\tau)\,\frac{d\tau}{\tau - Z}$$

for $|z| \neq 1$. This relation leads to a direct determination of the principal function.

Indeed, even if W is a partial isometry with $WV = (V + D)W$ the principal function of the dilated triplet $\{\tilde{W}, \tilde{V}, \tilde{D}\}$ can be obtained explicitly from knowledge of

$$\det[S_-(V;W - z)(\lambda)S_+(V;W - z)^{-1}(\lambda)]$$

for $|z| > 1$ and $\delta(\lambda)$. See [6].

There is also a nice relation between the various principal functions to which we will now turn.

Let $T \in \mathcal{m}$ with $[T^*,T] \in \mathcal{J}_T$, and define $T_z = T - aI$. Let the polar decomposition of T_z be given in the form $T_z = W_z(T_z^*T_z)^{1/2}$.

Let $g_z^p(\lambda, \ )$ be the principal function associated with the pair $(W_z, (T_z^*, T_z)^{1/2})$, and let $g_{(\nu,\mu)}$ be the principal function associated with $U = \dfrac{T + T^*}{2}$ and $V = \dfrac{T - T^*}{2i}$.

<u>Theorem 4.4.</u> Let $\delta + i\gamma = \sqrt{\lambda}\, e^{i\theta}$ for fixed $z = x + iy$. Then $g_z^p(\lambda, e^{i\theta}) = g(\gamma + y, \delta + x)dA$ a.e.

This result leads to a variety of theorems which can apparently not be obtained by other means. See for example [2] and [4]. We will not discuss it further here.

We will round out this survey by noting that we have (besides $E(\ell, z)$ and $\wp(\ell, \omega)$ and their associated functional calculi on the plane and on the cylinder) another determining function, ψ, with an associated functional calculus on the torus.

This determining function is appropriate for discussing the case of two unitary operators U and W with

$$U^*WUW^* - 1 \equiv \Gamma \in \mathcal{J}_T .$$

The analogs of theorems 1.1, 1.2, 2.2, 3.1, 3.2, 3.3 are valid as is an appropriate index result in terms of $g(e^{i\varphi}, e^{i\nu})$, the principal function which enters into the theorems which replace theorems 1.1 and 1.2 above.

Since the commutator relation written here for U and W is more general than the ones we have considered up to this point, the theory of the ψ determining function represents a increase in generality beyond the results described in more detail above.

The extension is, however, obtained by a straightforward process of reworking [3] with only minor changes and no surprises.

5. Symmetric Operators and Deficiency Indices.

Suppose $VU-UV = \frac{1}{\pi i} C$ and we define $W = (U+i)(U-i)^{-1}$. Then $WV = (V-D)W$ where $D = \frac{2}{\pi}(U-i)^{-1} C(U+i)^{-1}$ and, by Theorem 3.2, we have $g_{(W,V)}(\lambda, \frac{\nu+i}{\nu-i}) = g_{(U,V)}(\nu,\lambda)$.

We will study this fact in connection with the theory of certain symmetric singular integral operators.

The following paraphrases with simplifications results obtained in [12].

Suppose that $A(\lambda)$ is a measurable real valued function finite almost everywhere on (a,b) and that $k(\lambda)$ is square summable over (a,b).

Let us define a sequence of self-adjoint operators on $L^2(a,b)$ by setting for $n = 1,2,\ldots,$

$$L_n x(\lambda) = A_n(\lambda)x(\lambda) + \frac{1}{\pi i}\int_a^b \frac{\bar{k}_n(\lambda)k_n(\mu)}{(\mu-\lambda)} x(\mu)d\mu \tag{5.1}$$

where

$$A_n(\lambda) = \begin{cases} A(\lambda) & \text{if } |A(\lambda)| < n \\ 0 & \text{if } |A(\lambda)| > n \end{cases}$$

$$k_n(\lambda) = \begin{cases} k(\lambda) & \text{if } |k(\lambda)| < n \\ 0 & \text{if } |k(\lambda)| > n \end{cases}$$

Let $g_n(\nu,\lambda)$ be the characteristic function of the set of points (ν,λ) for which $A_n(\lambda) - |k_n(\lambda)|^2 < \nu < A_n(\lambda) + k_n(\lambda)|^2$.

Let $g(\nu,\lambda)$ be the characteristic function of the set of points (ν,λ) for which $A(\lambda) - |k(\lambda)|^2 < \nu < A(\lambda) + |k(\lambda)|^2$.

Define

$$h_\ell(t) = \begin{cases} \dfrac{\operatorname{Im} \ell}{\pi} \displaystyle\int_{-\infty}^{\infty} \frac{g(t,x)}{|x-\ell|^2}\, dx, & \operatorname{Im} \ell > 0 \\[2ex] \dfrac{|\operatorname{Im} \ell|}{\pi} \displaystyle\int_{-\infty}^{\infty} \frac{1-g(t,x)}{|x-\ell|^2}\, dx, & \operatorname{Im} \ell < 0 \end{cases} \tag{5.2}$$

With this definition we can see that

$$\begin{aligned} h_i(t) &= \frac{1}{\pi} \tan^{-1}[A(t) + |k(t)|^2] \\ &\quad - \frac{1}{\pi} \tan^{-1} [A(t) - |(t)|^2] . \end{aligned} \tag{5.3}$$

Since $0 \le h_\ell(t) \le 1$, we can verify that $\exp\{\int h_\ell(t) \frac{dt}{t - z}\}$ has imaginary part positive in the upper z plane.

Accordingly, there is a unique positive measure $d\mu_{\bar{\ell}}$ so that

$$\exp\{-\int_{-\infty}^{\infty} h_\ell(t) \frac{dt}{t - z}\} = 1 - \int \frac{d\mu_{\bar{\ell}}(t)}{t - z} \tag{5.4}$$

It is known from [13] that the sequence of operators $(L_n-\ell)^{-1}k_n \otimes (L_n-\ell)^{-1}k_n$ converges in trace norm to an operator of one dimensional range, D_ℓ, if and only if the measure $d\mu_\ell$ is absolutely continuous with respect to Lebesgue measure.

Let us simply assume now that $d\mu_{-i}$ is absolutely continuous. We will impose this condition throughout the following discussion.

It is known that the weak limit $\omega\text{-}\lim_{n\to\infty}(L_n-\ell)^{-1} = R(\ell)$ always exists and it becomes natural to ask for conditions under which $R(\ell)$ is the resolvent of a symmetric operator L (when ℓ is either in the upper or lower half plane).

We shall see that the condition $d\mu_{-i}$ is absolutely continuous, implies that $R(\ell) = (L-i)^{-1}$ for a certain maximal symmetric operator L, and we will compute the non-zero deficiency index of L in terms of the principal function $g(\nu,\lambda)$.

Suppose that we have $Vx(\lambda) = \lambda x(\lambda)$ so that

$$L_n V - VL_n = \frac{1}{\pi i} k_n \otimes k_n \quad (5.5)$$

Then if $W_n = (L_n+i)(L_n-i)^{-1}$ we can easily see that $W_n V = (V-D_n)W_n$ where

$$D_n = \frac{2}{\pi} (L_n-i)^{-1} C_n (L_n+i)^{-1} . \quad (5.6)$$

Assertion. If $d\mu_{-i}$ is absolutely continuous, then the sequence of unitary operators W_n converges strongly to an isometric operator W.

For the proof of this assertion, it suffices to note that $W_n = W_- \ f_n(V)$ where $f_n(V) = S_-(U;W_n)$ (f_n is some function since V has simple multiplicity and $W_-^{(n)}$ is the wave operator corresponding to the perturbation problem $V \to V-D_n$).

Let us note first, because D_n converges in trace norm, that a theorem of Kato [8] enables us to assert that the sequence $W_-^{(n)}$ is strongly convergent to W_-, the wave operator for $V \to V-D$.

Now we wish to show that the sequence $f_n(V)$ is strongly convergent. For this purpose, we will calculate $f_n(V)$ explicitly.

Form $S_-(V;W_n) = S_-(V;W_-^{(n)} f_n) = S_-(V;W_n^{(n)})S_-(V;f_n)$.

Since it is known [5] that $S_-(V;W_-^{(n)}) = P_a(V)$ we can con lude that

$$S_-(V;W_-^{(n)}) = f_n(V) .$$

But $W_n = (L_n+i)(L_n-i)^{-1}$. Hence

$$\begin{aligned} S_-(V;W_n) &= S_-(V;L_n+i)S_-(V;L_n-i)^{-1} \\ &= (A_n(\lambda) + |k_n(\lambda)|^2 + i)(A_n(\lambda) + |k_n(\lambda)|^2 - i)^{-1} . \end{aligned}$$

It is apparent that $f_n(V)$ is strongly convergent since $f_n(\lambda)$ converges in measure to

$$(A(\lambda) + |k(\lambda)|^2 + i)(A(\lambda) + |k(\lambda)|^2 - i) = f(\lambda) .$$

Thus, since multiplication is continuous in the strong topology, we can conclude that W_n is strongly convergent to $W_- f(V)$.

Furthermore, $WV = (V-D)W$.

Since the initial space of W_n is $\mathcal{H}_a(V) = \mathcal{H}$, we see that W is an isometry.

Indeed, it now follows that $W = (L+i)(L-i)^{-1}$ where L is a maximal symmetric operator. Further work shows that L coincides with the so-called weak graph limit of the sequence L_n (see, for example, [6] for the appropriate definition). Indeed, we have $R(i) = (L-i)^{-1}$.

It is easy to see now that $-h_i(t)$ is the spectral displacement function corresponding to the perturbation problem $V \to V-D$.

It follows from this observation that the defect of W is equal to the dimension of the singular space of $V-D$.

This dimension is easily computed to be $n = \dim L^2(\mu_i^s)$ where μ_i^s is the singular part of μ_i.

Thus the deficiency indices of L are (o,n) [12].

However much more can be said.

$$g_n(\nu,\lambda) = g_{(W_n V)}(\lambda,\frac{\nu-i}{\nu-i})$$

and, since $g_n(\nu,\lambda)$ is simply a truncation of $g(\nu,\lambda)$ and $\iint g(\nu,\lambda)d\nu d\lambda = \int_a^b |k(\lambda)|^2 d\lambda < \infty$, we can conclude that

$$\lim_{n\to\infty} g_n(\nu,\lambda) = g(\nu,\lambda) = g_{(W,V)}(\lambda,\frac{\nu+i}{\nu-i})$$

$$= \lim_{n\to\infty} g_{(W_n,V)}(\lambda,\frac{\nu+i}{\nu-i}) .$$

We can now go further. It is known that $g_{(W,V)}(\lambda,\frac{\nu+i}{\nu-i})$ is the complete unitary invariant of the pair $(\tilde{V},\tilde{W})$. This fact can be used, sometimes, to investigate the unitary part, if any, of $(L+i)(L-i)^{-1}$, [4; Theorem 10].

For example, if $a = 0$, $b = 1$ and $A(\lambda) = 0$, $k(\lambda) = \lambda^{-1/4}$, we can see at once that the deficiency indices of L are $(0,1)$ and that L is simple.

Given any $g(\nu,\mu)$ with $\iint g(\nu,\mu)d\nu d\mu < \infty$ and $0 \le g(\nu,\mu) \le 1$, there always xists [2] a pair of self-adjoint operators H_1 and H_2 with

$$(H_1-z_1)^{-1}(H_2-z_2)^{-1} - (H_2-z_2)^{-1}(H_1-z_1)^{-1}$$

$$= (H_2-z_2)^{-1}(\ _1-z_1)^{-1} \frac{C}{i}(H_1-z_1)^{-1}(H_2-z_2)^{-1}$$

and

$$\det(1 + \frac{1}{i} \hat{C}(H_1-\ell)^{-1}(H_2-z)^{-1} \hat{C})$$

$$= \exp \frac{1}{2\pi i} \iint g(\nu,\mu) \frac{d\nu}{\nu-\ell} \frac{d\mu}{\mu-t}$$

The known results from [12] which we have just surveyed correspond to the special case where $\tilde{W} = (H_1+i)(H_1-i)^{-1}$ and $R(\ell) = P(H_1-\ell)^{-1}$, $V = PH_2P$ with P being the projection onto the absolutely continuous part of H_2.

The study of relations between the operators H_1 and H_2 whose existence comes to us from the abstract determining function theory -- more specifically from symmetry and positivity properties of $E(\ell,z)$ -- and symmetric singular integral operators of arbitrary deficiency indices was one of the main original goals of the determining functions' theory.

<u>Note added in proof.</u>

The second author has now solved the very much more difficult problem of treating symmetric singular integral operators <u>without</u> the restriction -- assumed in the foregoing paragraphs -- that $d\mu_{-i}$ is absolutely continuous. The details will appear in [14].

It turns out that $1 + 2i\, R(i)$ restricted to a certain subspace -- which can be explicitly determined in terms of $g(\nu,\lambda)$ -- is a partial isometry and the Cayley transform of a symmetric opeartor L which coincides with the formal operator $A(\lambda)\ x\ (\lambda)$ + $\frac{1}{\pi i} \int \frac{\bar{k}(\lambda)k(a)}{\mu - \lambda}\ x\ (\mu)d\mu$ restricted to a dense domain of symmetry.

The domain of L is also explicitly obtained as are the

deficiency spaces. The deficiency indices of L are $n = \dim L_2(\mu_i^s)$, $m = \dim L_2(\nu_{-i}^s)$. All of the symmetric extensions of L can be described, and the method applies to symmetric Wiener-Hopf and Toeplitz operators. An associated model theory for restrictions of pairs of unitary operators is currently under study by the authors.

6. Mosaics.

We wish to sketch now the reason for the existence of the principal function.

First, consider the case where $\mathcal{M} = \mathcal{L}(\mathcal{H})$ and τ is the ordinary trace.

Define for fixed $\lambda \in sp(U)$, $\operatorname{Im} \ell \neq 0$

$$\theta(\lambda,\ell) = J + 2 \frac{d}{d\lambda} [\hat{C}^* E_\lambda (S_+(U;V) - \ell)^{-1}\hat{C}] .$$

$\theta(\lambda,\ell)$ is an operator valued analytic function with imaginary part positive in the upper half plane.

We can show that in fact

$$\theta(\lambda,\ell) = J\ E^*(\ell,\lambda + io) J\ E(\ell,\lambda + io) ,$$

and we have the following result.

<u>Theorem 6.1.</u> There exists a unique operator $B(\nu,\mu)$ in with $0 \leq B(\nu,\mu) \leq 1$ such that

$$\theta(\lambda,\ell) = \exp[i\pi P_- + \int \frac{B(\nu,\lambda)-P_-}{\nu-\ell} d\nu] .$$

Furthermore $\int \dfrac{B(\nu,\lambda)-P_-}{\nu-\ell} d\nu$, $P_+B(\nu,\lambda)P_+$, and $P_-(1-B(\nu,\lambda))P_-$ are all in trace class.

The function $B(\nu,\lambda)$ is called the mosaic of the pair (U,V).

<u>Theorem 6.2.</u>

$$g(\nu,\lambda) = \tau[P_+B(\nu,\lambda)P_+] - P_-[1-B(\nu,\lambda)]P_-] .$$

By this we mean that the $g(\nu,\lambda)$ given by the right hand side of this equation satisfies equation (1.3).

Remark 6.1. When $T-(\mu+i\nu)$ is Fredholm, $B(\nu,\lambda)$ is explicitly known. $B(\nu,\lambda)$ is the projection operator corresponding to $(-\infty,0)$ for the self-adjoint operator

$$J + 2 \frac{d}{d\lambda} \hat{C}^* E_\lambda (S_+(U;V)-\nu)^{-1} \hat{C} ,$$

and if P_- is finite, then

$$\text{Index } T-(\lambda+i\nu)1 = \dim \text{Range } B(\nu,u) - \dim \text{Range } P_- .$$

The situation is much more complex in the case where $\mathcal{M}$ is a type II_∞ factor.

In this case the boundary behavior of $E(\ell,z)$ can be very bad, and local data may not exist.

Nevertheless, a more delicate argument using the exponential representation which we have found for operator valued functions with imaginary part positive in the upper half plane suffices to show the existence of a principal function.

We will sketch these considerations. The reader is referred to [3] for the complete proofs.

Form

$$\theta(\ell,z) = E^*(\bar{\ell},z)JE(\ell,z) = J+k_z(V-\ell)^{-1}k_z^*$$

with $k_z \equiv \sqrt{-i(z-\bar{z}}\, \hat{C}^*(U-\bar{z})^{-1}$.

There exists an element $B(\cdot,z) \in L^\infty(R^1,dt,\mathcal{M})$ with $0 \le B(t,z) \le 1$ and

$$\theta(\ell,z) = \exp\{i\pi P_- + \int \frac{B(t,z)-P_-}{t-\ell}\, dt\}$$

as long as z is a fixed point in the upper half plane.

For fixed $\eta > 0$ form the map $B:R^1 \to L^\infty(R^1,dt,\mathcal{M})$ given by $\lambda \to B(\cdot,\lambda+i\eta)$.

There is a product measurable representation of this map, i.e. there is a unique element B_η of $L^\infty(R^2,dA,\mathcal{M})$ such that for almost all λ,

$$B_\eta(t,\lambda) = B(t,\lambda+i\eta) .$$

This family lies in the unit sphere of $L^{\infty}(R^2,dA,\mathfrak{M})$ which is weakly compact. Since $\mathfrak{M}_*$ is separable and $L^1(R^2,dA,\mathfrak{M}_*)^*=L^{\infty}(R^2,dA,\mathfrak{M})$ we can select a positive sequence $\eta_j \to 0$ so that $B_{\eta_n}(\cdot,\cdot)$ converges weakly in $L^{\infty}(R^2,dA,\mathfrak{M})$ to a limit $B(\cdot,\cdot)$. From this we can show that the sequence $g_{\eta}(\nu,\lambda) = \tau(P_+B_{\eta}(\nu,\lambda)P_+ - P_-(1-B(\nu,\lambda))P_-)$ also turns out to converge in the $L^1(dA)$ norm to

$$g(\nu,\lambda) = \tau(P_+B(\nu,\lambda)P_+ - P_-(1-B(\nu,\lambda))P_-) .$$

This $g(\nu,\lambda)$ satisfies (1.3) -- and therefore does not depend upon the choice of B_{η} subsequence!

7. A Trace Formula for Multi-generators.

In this section we shall develop a Poisson-Bracket formula for type I_{∞} almost commuting C*-algebras which have a finite number of self-adjoint generators $X_1,X_2,\ldots,X_n,Y$ with the property that $[X_i,X_j] = 0$ for $1 \le i,j \le n$. The formula, which we state only for the case $n = 2$, is expressed as a double integral whose integrand consists of "weighted" partial Jacobians coming from the spectral representation of each of the operators X_1 and X_2. The effect of having $n = 2$ leads to a decomposition of the pairwise principal functions $g_j(\nu,\lambda)$ associated with $\{X_j,Y\}$, $j = 1,2$, into partial principal functions ("partial indices") $g_{ij}(\nu,\alpha(\lambda);\lambda)$, $i \neq j$, whose sum taken over a natural indexing set equals $g_j(\nu,\lambda)$. These relations are of course invariant under trace class perturbations of X_1, X_2 and Y. Examples of such operators occur in C*-algebras on a half-space of Z^n.

In as much as the Poisson-Bracket formula for a pair of operators provides for an index formula relating traces, index and homology, the present formula has an analogous interpretation.

As indicated, we shall restrict our attention to the case where $n = 2$, the arguments for larger values of n being straightforward.

Let $M(R^3)$ be the space of complex measures w over R^3 such that

$$\|w\| = \int_{R^3}(1+|t|)(1+|s|)(1+|r|)d|w(t,s,r)| < \infty .$$

Let $\hat{M}(R^3)$ be the set of all characteristic functions of measures in $M(R^3)$. We can associate with $F \in \hat{M}(R^3)$ and the triplet $\{Y_1, X_2, Y\}$ an element in the C*-algebra A generated by X_1, X_2 and Y by the iterated triple integral

$$F(X_1,X_2,Y) = \iiint F(x_1,x_2,y)dE_{x_1}\, dF_{x_2}\, dG_y$$

where E_{x_1}, F_{x_2} and G_y are the respective spectral resolutions of X_1, X_2 and Y. The association $F \to F(X_1,X_2,Y)$ defines a functional calculus of the Mikhlin-Weyl type.

Now, let F and $H \in \hat{M}(R^3)$. We seek a representation of

$$\tau[F(X_1,X_2,Y),H(X_1,X_2,Y)]$$

in terms of the functions F and H. To this end we introduce linear functionals ℓ_1 and ℓ_2 (corresponding to X_1 and X_2) by setting

$$(71.) \qquad \ell_1\left(\frac{\partial H}{\partial Y}\right) = \tau[X_1,H(X_1,X_2,Y)]$$

and

$$(7.2) \qquad \ell_2\left(\frac{\partial H}{\partial Y}\right) = \tau[X_2H(X_1,X_2,Y)] .$$

Since X_1 and X_2 commute, the functionals ℓ_1 and ℓ_2 are well defined on $\hat{M}(R^3)$ and are invariant under trace class perturbations of X_1, X_2 and Y.

Let $J_{13}(F,H)$ and $J_{23}(F,H)$ denote the partial Jacobians

$$\frac{\partial F}{\partial X_1}\frac{\partial H}{\partial Y} - \frac{\partial F}{\partial Y}\frac{\partial H}{\partial X_1}$$

and

$$\frac{\partial F}{\partial X_2}\frac{\partial H}{\partial Y} - \frac{\partial F}{\partial Y}\frac{\partial H}{\partial X_2}$$

respectively. We claim that

$$(7.3)\qquad \tau[F(X_1,X_2,Y),H(X_1,X_2,Y)] = \ell_1(J_{13}(F,H)) + \ell_2(J_{23}(F,H)).$$

To prove this, consider the bilinear form

$$(F,H) \to \langle F,H\rangle = \tau[F(X_1,X_2,Y),H(X_1,X_2,Y)]$$
$$- \ell_1(J_{13}(F,H)) - \ell_2(J_{23}(F,H)) \ .$$

We first observe that

$$\langle x_1,H\rangle = \tau[X_1,H(X_1,X_2,Y)] - \ell_1(\tfrac{\partial H}{\partial Y}) = 0$$

and

$$\langle x_2,H\rangle = \tau[X_2,H(X_1,X_2,Y)] - \ell_2(\tfrac{\partial H}{\partial Y}) = 0 \ .$$

Now recall that the radical of the bilinear form $\langle,\rangle$ is the set $\{H|\langle F,H\rangle = 0,\ F \in \hat{\mathfrak{M}}(R^3)\}$ and observe that $\langle,\rangle$ has the collapsing property in the sense that, if $P = F\circ H$ and $Q = G\circ H$ where F and G are functions of a single variable, then $\langle P,Q\rangle = 0$. Now the collapsing property of $\langle,\rangle$ implies that its radical is an algebra [7]. Thus $\langle F(X_1,X_2,Y),H(X_1,X_2,Y)\rangle = 0$ for F,H in $\hat{\mathfrak{M}}(R^3)$ and $\frac{\partial F}{\partial Y} = 0$. Collapsing again implies that Y also belongs to the radical [7]. Thus $\langle F,H\rangle = 0$ for all F and H. This establishes the validity of (7.3).

We now seek an expression for the functionals ℓ_1 and ℓ_2. Here is where we must restrict ourselves to the type I_∞ case. The idea is to observe that for arbitrary H in $\hat{\mathfrak{M}}(R^3)$

$$(7.4)\qquad \tau[X_1,H(X_1,X_2,Y)] = \frac{i}{2\pi}\int_{sp(X_1)} \tau\{S_-(X_1;H(X_1,X_2,Y))(\lambda) - S_+(X_1;H(X_1,X_2,Y)(\lambda)\}d\lambda$$

and

$$(7.5)\qquad \tau[X_2,H(X_1,X_2,Y)] = \frac{i}{2\pi}\int_{sp(X_2)} \tau\{S_-(X_2;H(X_1,X_2,Y))(\lambda) - S_+(X_2;H(X_1,X_2,Y))(\lambda)\}d\lambda \ .$$

These relations follow almost immediately from the known formulae for the case of a pair of operators.

We shall work below with the first expression above, that is with symbols taken relative to X_1, similar considerations being evident for symbols relative to X_2.

The homomorphism property of the symbol maps implies

$$(7.6) \qquad S_+(X_1;H(X_1,X_2,Y))(\lambda) = H(\lambda, S_{\pm}(X_1;X_2)(\lambda), S_{\pm}(X_1;Y)(\lambda))$$

as operators on an appropriate Hilbert space $\mathscr{K}(\lambda)$ in the direct integral diagonalizing space for X_1.

But $S_+(X_1;X_2)(\lambda) = S_-(X_1;X_2)(\lambda) \equiv S(X_1,X_2)(\lambda)$, since X_1 and X_2 commute. Also, since $[X_2,Y]$ is trace class, the symbols of X_2 and Y relative to X_1 also commute. Thus, $S(X_1;X_2)(\lambda)$ commutes with both $S_+(X_1;Y)(\lambda)$ and $S_-(X_1;Y)(\lambda)$.

For fixed λ let $P_d(\lambda)$ denote the projection of $\mathscr{K}(\lambda)$ onto the closed subspace spanned by the eigenvectors of $S(X_1;X_2)(\lambda)$. Then $P_d(\lambda)$ commutes with $S_+(X_1:Y)(\lambda)$ and $S_-(X_1;Y)(\lambda)$. Moreover, since $S_+(X_1;Y)(\lambda) - X_-(X_1;Y)(\lambda)$ is trace class on $\mathscr{K}(\lambda)$ [5], $S_+(X_1;Y)(\lambda)f = S_-(X_1;Y)(\lambda)f$ for f in the kernel of $P_d(\lambda)$. This last statement follows from the fact that if a compact operator T lies in the commutant of a self-adjoint operator having purely continuous spectra, then T just be the zero operator.

Consequently, in order to evaluate the integrand in (7.4), it suffices to assume that the eigenvectors of $S(X_1;X_2)(\lambda)$ span $\mathscr{K}(\lambda)$.

Let $\Sigma\alpha(\lambda)E_{12}[\alpha(\lambda);\lambda]$ denote the spectral resolution of $S(X_1;X_2)(\lambda)$ where the summation is extended over all points $\alpha(\lambda)$ belonging to $\Delta_{12}(\lambda)$, the point spectrum of $S(X_1;X_2)(\lambda)$, and $E_{12}[\alpha(\lambda);\lambda]$ denotes the corresponding eigenprojector. Since each projection $E_{12}[\alpha(\lambda);\lambda]$ commutes with $S_+(X_1;Y)(\lambda)$ and $S_-(X_1;Y)(\lambda)$ we evidently have

(7.7)

$$\tau\{H(\lambda,S(X_1;X_2)(\lambda),S_-(X_1;Y)(\lambda)) -$$

$$- H(\lambda,S(X_1;X_2(\lambda),S_+(X_1;Y)(\lambda))\} =$$

$$\sum_{\alpha(\lambda)\in\Delta_{12}(\lambda)} \tau\{H(\lambda,\alpha(\lambda),S_-(X_1;Y)(\lambda)E_{12}[\alpha(\lambda);\lambda])$$

$$- H(\lambda,\alpha(\lambda),S_+(X_1;Y)(\lambda)E_{12}[\alpha(\lambda);\lambda])\} =$$

$$\sum_{\alpha(\lambda)\in\Delta_{12}(\lambda)} \int \frac{\partial H}{\partial Y}(\lambda,\alpha(\lambda),\nu)g_{12}(\nu.\alpha(\lambda);\lambda d\nu$$

where $\nu \rightarrow g_{12}(\nu,\alpha(\lambda);\lambda)$ is the phase shift corresponding to the "cut-down" perturbation

$$S_+(X_1;X_2)(\lambda)E_{12}[\alpha(\lambda);\lambda] \rightarrow S_-(X_1;X_2)(\lambda)E_{12}[\alpha(\lambda);\lambda] .$$

Insertion of these last identities into (7.4) yields

(7.8)

$$\ell_1(\frac{\partial H}{\partial Y}) = \tau[X_1,H(X_1,X_2,Y)]$$

$$= \iint\{ \sum_{\alpha(\lambda)\in\Delta_{12}(\lambda)} \frac{\partial H}{\partial Y}(\lambda,\alpha(\lambda),\nu)g_{12}(\nu,\alpha(\lambda);\lambda)\}d\nu d\lambda .$$

Similarly, with $\Delta_{21}(\lambda)$ denoting the point spectrum of $S_+(X_2;X_1)(\lambda) = S_-(X_2;X_1)(\lambda)$ (here $S_\pm(X_2;X_1)(\lambda)$ denotes the pointwise representation of $S_\pm(X_2;X_1)$ in a direct integral diagonalizing space for X_2) there is a corresponding family of partial indices $g_{21}(\nu,\alpha(\lambda);\lambda)$ such that

(7.9)

$$\ell_2(\frac{\partial H}{\partial Y}) = \tau[X_2,H(X_1,X_2,Y)]$$

$$= \frac{i}{2\pi}\iint \sum_{\alpha(\lambda)\in\Delta_{21}(\lambda)} \frac{\partial H}{\partial Y}(\alpha(\lambda),\lambda,\nu)g_{21}(\nu,\alpha(\lambda);\lambda)d\nu d\lambda.$$

In each case it is clear that

$$\sum_{\alpha(\lambda)\in\Delta_{12}(\lambda)} g_{12}(\nu,\alpha(\lambda),\lambda) = g_1(\nu,\lambda)$$

and

$$\sum_{\alpha(\lambda)\in\Delta_{21}(\lambda)} g_{21}(\nu,\alpha(\lambda);\lambda) = g_2(\nu,\lambda)$$

where $g_1(\nu,\lambda)$ and $g_2(\nu,\lambda)$ are the principal functions for the pairs $\{X_1,Y\}$ and $\{X_2,Y\}$, repsectively. If $\lambda \to \alpha(\lambda)$ and $\lambda \to E_{12}[\alpha(\lambda);\lambda]$ are chosen to be measurable, then the function $(\nu,\lambda) \to g_{12}(\nu,\alpha(\lambda);\lambda)$ is the principal function for the pair $\{X_1,Y\}$ "compressed" to the subspace which is the range of the projection $\int \oplus E_{12}[\alpha(\lambda);\lambda]d\lambda$.

Combining (7.3) with (7.8) and (7.9) we have the following representation

Theorem 7.1. Let X_1,X_2 and Y be the self-adjoint operators with $[X_1,X_2] = 0$, $[X_1,Y]$ and $[X_2,Y]$ trace class. Let

$$\sum_{\alpha(\lambda)\in\Delta_{12}(\lambda)} \alpha(\lambda)E_{12}[\alpha(\lambda);\lambda]$$

and

$$\sum_{\alpha(\lambda)\in\Delta_{21}(\lambda)} \alpha(\lambda)E_{21}[\alpha(\lambda);\lambda]$$

denote the spectral resolutions of $S_{\pm}(X_1;X_2)(\lambda)$ and $S_{\pm}(X_2;X_1)(\lambda)$ and $S_{\pm}(X_2;X_1)$ in direct integral Hilbert spaces in which X_1 and X_2 are diagonalized. Then, with $g_{12}(\nu,\alpha(\lambda);\lambda)$ and $g_{21}(\nu,\alpha(\lambda);\lambda)$ denoting the phase shifts corresponding to the respective cut down perturbations

$$S_+(X_1;X_2)(\lambda)E_{12}[\alpha(\lambda);\lambda] \to S_-(X_1;X_2)(\lambda)E_{12}[\alpha(\lambda);\lambda]$$

$$S_+(X_2;X_1)(\lambda)E_{21}[\alpha(\lambda);\lambda] \to S_-(X_2;X_1)(\lambda)E_{21}[\alpha(\lambda);\lambda]$$

for any F,H in $\hat{M}(R^3)$

$$\tau[F(X_1,X_2,Y,H(X_1,X_2,Y)]$$

$$= \frac{i}{2\pi}\iint \sum_{\alpha(\lambda)\in\Delta_{12}(\lambda)} J_{13}(F.H)(\lambda,\alpha(\lambda),\nu)g_{12}(\nu,\alpha(\lambda);\lambda)$$

$$+ \sum_{\alpha(\lambda)\in\Delta_{21}(\lambda)} J_{23}(F,H)(\alpha(\lambda),\lambda,\nu)g_{21}(\nu,\alpha(\lambda);\lambda)\}d\nu d\lambda .$$

Using this formula it would be interesting to explore various transformation properties. For instance if $F, H \in \hat{M}(R^3)$ and are real-valued, it should be possible to express the principal function of the operator $F(X_1, X_2, Y) + iH(X_1, X_2, Y)$ in terms of the partial $g_{12}(\nu, \alpha(\lambda); \lambda)$ and $g_{21}(\nu, \alpha(\lambda); \lambda)$.

Consider the case of the triplet of operators $\{X_1, X_2, W\}$ where W is unitary X_1 and X_2 self-adjoint, X_1 and X_2 commute and both $[W, X_1]$, $[W, X_2]$ are trace class.

The formula in Theorem 5.1 may not be valid if sp(W) is the full circle i.e. the functionals ℓ_1 and ℓ_2 are not sufficient. This reflects the fact that the principal function corresponding to an intertwining triplet is not determined by the symbols alone [6].

REFERENCES

1. M. Breuer, Fredholm theories in von Neumann algebras I and II, Math. Ann. 178 (1968), 243-254 and 180 (1969), 313-325.

2. R. Carey and J. Pincus, Construction of seminormal operators with prescribed mosaic, Indiana Univ. Math. J. 23 (1974), 1155-1165.

3. ______, Mosaics, principal functions and mean motion in von Neumann algebras, to appear in Acta Mathematica.

4. ______, An invariant for certain operator algebras, Proc. Nat. Acad. Sci. 71 (1974), 1952-1956.

5. ______, Commutators, symbols and determining functions, J. Functional Analysis, 19 (1975), 50-80.

6. ______, The structure of intertwining partial isometries II, to appear.

7. J. W. Helton and R. E. Howe, (1973), Integral operators, commutator traces, index and homology, Proc. of a conference on operator theory, Springer-Verlag Lecture Notes No. 345.

8. T. Kato, Perturbation Theory for Linear Operators, Springer-Verlag, New York, 1966.

9. M. G Krein, Perturbation determinants and a formula for traces of unitary and self-adjoint operators, Dokl. Akad. Nauk. SSSR 144 (1962), 268-271.

10. J. D. Pincus, Commutators and systems of singular integral equations, I, Acta. Math. 121 (1968), 219-249.

11. ___________, On the trace of commutators in the algebra of operators generated by an operator with trace class self-commutator, unpublished preprint, dated August, 1972.

12. ___________, Symmetric singular integral operators, Indiana Univ. Math. J. 23 (1973), 537-556.

13. J. D. Pincus and J. Rovnyak, A spectral theory for some unbounded self-adjoint singular integral operators, Amer. J. Math. 91 (1969) 619-636.

14. J. D. Pincus, Symmetric singular integral operators with arbitrary deficiency, to appear in Advances in Mathematics, .M.G. Krein Anniversary Issue).

15. ___________, The determining function method in the treatment of commutator systems, Colloquia Math. Soc., Janos Bolyai5, Hilbert Space Operators, 1970.

EXTENSIONS OF C^*-ALGEBRAS AND K-HOMOLOGY*

R.G. Douglas, State University of New York at Stony Brook

In this talk I shall describe my recent joint work with L.G. Brown and P.A. Fillmore. Since the connections of the work with operators have been explored in some detail in [8], I shall emphasize some other aspects of the work and, in particular, its relation to algebraic topology. Details of the results announced in [7] will be given in [9] as well as in various papers now in preparation. Additional discussion of this topic is contained in the talk of L.G. Brown in this volume [6].

I begin by fixing some notation. For $\mathcal{H}$ a separable complex Hilbert space, let $\mathcal{L}$ $(=\mathcal{L}(\mathcal{H}))$ denote the C^*-algebra of all bounded linear operators on $\mathcal{H}$, $\mathcal{K}(=\mathcal{K}(\mathcal{H}))$ the two-sided ideal of compact operators on $\mathcal{H}$, $\mathcal{Q}(=\mathcal{Q}(\mathcal{H}))$ the quotient C^*-algebra $\mathcal{L}/\mathcal{K}$, and π the natural $*$-homomorphism from $\mathcal{L}(\mathcal{H})$ onto $\mathcal{Q}(\mathcal{H})$. For X a compact metric space let C(X) be the commutative C^*-algebra of continuous complex-valued functions on X. One knows that C(X) is the most general separable commutative C^*-algebra.

By an extension of $\mathcal{K}(\mathcal{H})$ by C(X) we shall mean a pair $(\mathcal{E},\varphi)$, where $\mathcal{E}$ is a C^*-subalgebra of $\mathcal{L}(\mathcal{H})$ containing $\mathcal{K}(\mathcal{H})$ and the identity operator $I_{\mathcal{H}}$ and φ is a $*$-homomorphism from $\mathcal{E}$ onto C(X) with kernel $\mathcal{K}(\mathcal{H})$. Such a pair yields the short exact sequence:

$$0 \longrightarrow \mathcal{K}(\mathcal{H}) \xrightarrow{\ i\ } \mathcal{E} \xrightarrow{\ \varphi\ } C(X) \longrightarrow 0,$$

where i is the inclusion map. (For the precise relation of this definition to what the usual algebraic definition would yield in the category of C^*-algebras, see [8], §4.) As might be expected one is not interested in the set of all extensions but rather a certain set of equivalence classes of extensions. The relevant notion of equivalence

*Research supported in part by a grant from the National Science Foundation.

is defined by $(\mathcal{E}_1,\varphi_1) \sim (\mathcal{E}_2,\varphi_2)$ if and only if there exists a *-isomorphism α from $\mathcal{E}_1$ onto $\mathcal{E}_2$ such that the following diagram commutes:

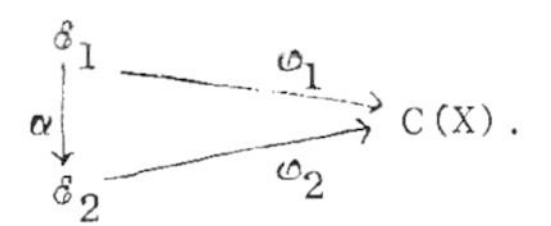

(This is in contrast with the usual definition of equivalence used in homological algebra, in which one requires that α induce the identity automorphism on $\mathcal{K}$.) Let Ext(X) denote the set of all equivalence classes of extensions of $\mathcal{K}$ by C(X). Before continuing, let us describe some contexts in which extensions arise.

If T is an essentially normal operator on $\mathcal{H}$, that is $[T,T^*] = TT^* - T^*T$ is compact, then one obtains the extension $(\mathcal{E}_T,\varphi_T)$, where $\mathcal{E}_T$ is the C*-subalgebra of $\mathcal{L}(\mathcal{H})$ generated by $I_{\mathcal{H}}$, T and $\mathcal{K}(\mathcal{H})$, and φ_T is the *-homomorphism from $\mathcal{E}_T$ onto $C(\sigma_e(T))$ satisfying $\varphi_T(T) = z$, where $\sigma_e(T)$ is the essential spectrum of T and z is the identity function on $\sigma_e(T) \subset \mathbb{C}$. For X a subset of $\mathbb{C}$ it is easy to see that all extensions of $\mathcal{K}$ by C(X) arise in this manner and determining Ext(X) is equivalent to classifying the essentially normal operators up to unitary equivalence modulo $\mathcal{K}$. In an entirely analogous manner one obtains the extensions for X a subset of $\mathbb{C}^n$ by considering n-tuples of operators $(T_1,T_2,\ldots T_n)$ on $\mathcal{H}$ such that $[T_i,T_j]$ and $[T_i,T_j^*]$ are compact for all i, j = 1,2,...n

In various contexts the Toeplitz operators with continuous symbol define an extension. More precisely, if R is an open subset of $\mathbb{C}$ with boundary ∂R consisting of a finite number of simple closed Jordan curves, then the C*-algebra generated by the Toeplitz operators with continuous symbol defined on the associated Hardy space gives rise to an element in Ext(∂R) [1], [11]. Further, if G is a strongly pseudoconvex domain in $\mathbb{C}^n$, then the C*-algebra generated by the Toeplitz operators with continuous symbol defined on the Bergman space of holomorphic functions on G relative to volume measure defines an element of Ext(∂G) [10], [14]. The classification of these extensions is closely related to the index problem for this class of operators [11] which remains open for general G in $\mathbb{C}^n$.

If M is a compact differentiable manifold without boundary, then the C*-algebra generated by the zeroth order pseudo-differential operators with scalar coefficients together with the compact operators on the L^2 space of some smooth volume measure defines an element in Ext(Sph*(M)), where Sph*(M) denotes the sphere bundle for the cotangent bundle of M. The study of this extension is closely related to the Atiyah-Singer index theorem and might be useful in providing analytical proofs for some deep theorems in topology [13]. For example, it can be shown that this extension distinguishes some of the different differentiable structures on a manifold.

Lastly, if τ is a *-monomorphism of C(X) into the Calkin algebra $\mathcal{Q}$, then one obtains an element of Ext(X) by setting $\mathfrak{E} = \pi^{-1}(\text{im } \tau)$ and $\varphi = \tau^{-1} \circ \pi$. Actually, this example provides a useful alternate definition of extensions, since $\tau = \pi \circ \varphi^{-1}$ is a well-defined *-monomorphism from C(X) into $\mathcal{Q}$ for an extension $(\mathfrak{E}, \varphi)$. Two *-monomorphism τ_1 and τ_2 define equivalent extensions if and only if there exists a unitary operator U on $\mathcal{H}$ such that $\tau_1 = \pi(U)^* \tau_2 \pi(U)$. I shall use this definition for the elements of Ext(X) interchangeably in what follows with the one given previously.

One of the most important general facts about Ext(X) is that it is an abelian group. Addition on Ext(X) is defined as follows. For *-monomorphisms τ_1 and τ_2 from C(X) into $\mathcal{Q}(\mathcal{H})$ we define $\tau_1 + \tau_2$ to be the *-monomorphism from C(X) to $\mathcal{Q}(\mathcal{H} \oplus \mathcal{H})$ such that

$$(\tau_1 + \tau_2)(f) = \tau_1(f) \oplus \tau_2(f) \quad \text{for } f \text{ in } C(X).$$

One must, of course, show that this is well-defined; it is, and Ext(X) becomes a commutative semigroup. A *-monomorphism τ from C(X) into $\mathcal{Q}(\mathcal{H})$, is said to define a trivial extension if τ lifts to a

$*$-monomorphism σ from $C(X)$ into $\mathcal{L}(\mathcal{H})$, that is, there exists σ such that the following diagram commutes:

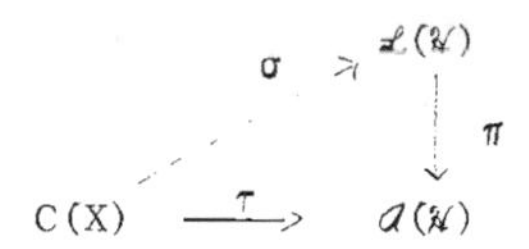

In [8] we prove that there exists a unique (to within equivalence) trivial extension for each X which is a unit in the semigroup Ext(X). Although the proofs are not applicable, the idea of making algebra extensions into a commutative semigroup with unit is familiar in homological algebra. That one should obtain a group, however, is quite unexpected. Our original proof of the fact was quite indirect and came nearly at the end of [8]. A direct proof has since been given by Arveson [4] using a dilation theorem due to Stinespring and a recent result on the existence of cross-sections for $*$-epimorphisms of Vesterstrøm [15] and Andersen [2] (see also Ando [3]). A simple proof of the special case of the latter result needed by Arveson has been given by Davie (cf. [9]).

Thus $X \longmapsto \mathrm{Ext}(X)$ defines a correspondence from compact metric spaces to abelian groups and a natural question is how does Ext(X) depend on X? If φ is a continuous map from X to Y, then φ induces the $*$-homomorphism $\varphi^*: C(Y) \longrightarrow C(X)$ defined by $\varphi^* f = f \circ \varphi$ for f in C(Y). If τ is a $*$-monomorphism from $C(X)$ into $\mathcal{Q}(\mathcal{H})$, then $\varphi_*(\tau) = \tau \circ \varphi^* + \tau_Y$ is a $*$-monomorphism from $C(X)$ into $\mathcal{Q}(\mathcal{H} \oplus \mathcal{H}')$, where τ_Y is a $*$-monomorphism from $C(X)$ into $\mathcal{Q}(\mathcal{H}')$ which defines the trivial extension. One can easily show that the correspondence $X \longmapsto \mathrm{Ext}(X)$ and $\varphi \longmapsto \varphi_*$ defines a covariant functor from the category of compact metric spaces and continuous maps to the category of abelian groups and homomorphisms.

In determining the group Ext for a particular space, one tries to reduce the calculation to simpler spaces. A key ingredient in the program is a Mayer-Vietoris sequence for Ext. One exploits the spectral theorem to enlarge im τ to a commutative C*-algebra with a simpler maximal ideal space. Doing this we obtain that the sequence

$$\mathrm{Ext}(A) \longrightarrow \mathrm{Ext}(X) \longrightarrow \mathrm{Ext}(X/A)$$

is exact for A a closed subset of X and that the sequence

$$\mathrm{Ext}(A) \xrightarrow{\alpha} \mathrm{Ext}(B) \oplus \mathrm{Ext}(C) \xrightarrow{\beta} \mathrm{Ext}(X)$$

is exact for A, B and C closed subsets of X such that $X = B \cup C$ and $A = B \cap C$, where $\alpha = i_{B*} \oplus i_{C*}$ for i_B and i_C the inclusion maps of A in B and C, respectively, and $\beta = j_{B*} - j_{C*}$ for j_B and j_C the inclusions maps of B and C in X. Further development of these ideas eventually leads to a proof that Ext of the cone over a space is trivial and hence Ext is a homotopy invariant functor. Thus we can apply standard techniques from algebraic topology to define a generalized homology theory based on Ext.

To this end we define $\mathrm{Ext}_1(X) = \mathrm{Ext}(X)$, $\mathrm{Ext}_0(X) = \mathrm{Ext}(SX)$ and $\mathrm{Ext}_{-1}(X) = \mathrm{Ext}(S^2X)$, where SX and S^2X are the first and second suspensions of X. We could of course define $\mathrm{Ext}_k(X)$ for all $k \leq +1$ but it is unnecessary. We show that there exists a natural isomorphism $\mathrm{Per}:\mathrm{Ext}(S^2X) \longrightarrow \mathrm{Ext}(X)$ for each compact metric space X and hence we can define $\mathrm{Ext}_k(X) = \begin{cases} \mathrm{Ext}(X) & k \text{ odd} \\ \mathrm{Ext}(SX) & k \text{ even} \end{cases}$. Our result establishes Bott periodicity for the generalized homology theory defined by Ext.

Again from general principles one establishes the existence of the long exact sequence, the long Mayer-Vietoris sequence, and indeed all the axioms of Eilenberg and Steenrod for a reduced homology theory except the dimension axiom. In fact, one shows more -- namely Ext defines what Milnor has called a Steenrod homology. Further, one can show that the homology theory defined by Ext corresponds to the cohomology theory defined by K-theory. Brown will have more to say about these matters in his talk. We shall consider some of the pairing which would have to exist for such a relation to be valid.

We begin by exhibiting a homomorphism γ_∞ from $\mathrm{Ext}_1(X)$ to $\mathrm{Hom}(K^1(X),Z)$. Recall that the set of Fredholm operators on $\mathcal{H}$ is precisely the inverse image under π of the group $\mathcal{Q}^{-1}$ of invertible elements in $\mathcal{Q}$ and that index defines a homomorphism ind from $\mathcal{Q}^{-1}$ to Z. Now let τ be a fixed $*$-monomorphism from C(X) to $\mathcal{Q}(\mathcal{H})$. If we consider the homomorphism $\mathrm{ind}\circ\tau$ from the group of invertible functions $C(X)^{-1}$ in C(X) to Z, then it is continuous and hence induces a homomorphism $\gamma_1(\tau)$ to Z with domain the group $[X,\mathbb{C}^*]$ of homotopy classes of maps from X to the non-zero complex numbers $\mathbb{C}^*$. Similarly, for each integer n τ defines a $*$-monomorphism τ_n from $C(X)\otimes M_n \cong C_{M_n}(X)$ to $\mathcal{Q}(\mathcal{H})\otimes M_n \cong \mathcal{Q}(\mathcal{H}\otimes\mathbb{C}^n)$, where M_n is the C*-algebra on n x n matrices on $\mathbb{C}^n$ and $C_{M_n}(X)$ is the C*-algebra of continuous M_n-valued functions on X. As before we can define the homomorphism $\mathrm{ind}\circ\tau_n$ from $C_{Gl_n}(X) = C_{M_n}(X)^{-1}$ to Z, where Gl_n is the group of invertible matrices on $\mathbb{C}^n$, and then induces a homomorphism $\gamma_n(\tau)$ to Z with domain the group $[X,Gl_n]$ of homotopy classes of maps from X to Gl_n. If we map Gl_n to Gl_{n+1} such that A in Gl_n is mapped to $\begin{pmatrix}A & 0\\ 0 & 1\end{pmatrix}$, then the following diagram commutes:

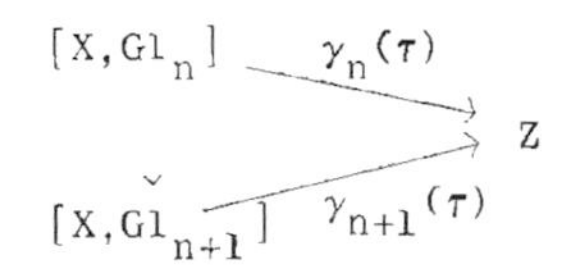

Hence we can define $\gamma_\infty(\tau)$ from $K^1(X) = \varinjlim[X,Gl_n]$ to Z. One must check that γ_∞ is a homomorphism on $\mathrm{Ext}_1(X)$, and that γ_∞ is functorial. For X homeomorphic to a subspace of R^3, we show in [9] that γ_∞ is an isomorphism and this is the essence of the classification of essentially normal operators which we obtain in [8]. For more general spaces γ_∞ is not an isomorphism, in part, because it kills the torsion subgroup of $\mathrm{Ext}_1(X)$. The map γ_∞ is always onto and its kernel can be identified, (cf. [6]).

Let us describe another pairing between Ext and K-theory which has interesting consequences for the theory of C*-algebras. We have already mentioned the C*-algebra $M_n(X)$; if one considers the extensions of $\mathcal{K}$ by $M_n(X)$, then one can show without too much difficulty that relative to a slightly weaker notion of equivalence, the corresponding equivalence classes form an abelian group naturally isomorphic to Ext(X). Moreover, this holds more generally. Let E be a rank n Hermitian vector bundle over X, that is, E is a topological space together with a continuous map ρ from E onto X such that the fiber $\rho^{-1}(x)$ over each x in X is an n-dimensional complex Hilbert space and such that E is locally a product space. A continuous map f on E is said to be a bundle map if f is a linear transformation on each fiber. The collection of bundle maps for E form a C*-algebra which we denote by $C_E(X)$ which for the trivial bundle $E = X \times \mathbb{C}^n$ yields $C_{X \times \mathbb{C}^n}(X) = C_{M_n}(X)$. Again the collection of "weak equivalence classes" of extensions of $\mathcal{K}$ by $C_E(X)$ can be shown to form an abelian group $\mathrm{Ext}(C_E(X))$ which is naturally isomorphic to Ext(X) and we let ξ_E denote the isomorphism from Ext(X) onto $\mathrm{Ext}(C_E(X))$. An element in $\mathrm{Ext}(C_E(X))$ corresponds to a *-monomorphism ρ from $C_E(X)$ into $\mathcal{Q}$; the center Z of $C_E(X)$ is naturally isomorphic to C(X) and hence if we restrict τ to Z we obtain an element of Ext(X) which we denote by $\zeta_E(\rho)$. By composing $\zeta_E \circ \xi_E$ we obtain an endomorphism μ_E on Ext(X) which for $E = X \times \mathbb{C}^n$ is just multiplication by n. The correspondence $E \longrightarrow \mu_E$ can be shown to extend to a map from the Grothendieck ring $K^0(X)$ into $\mathrm{End}(\mathrm{Ext}_1(X))$ which makes $\mathrm{Ext}_1(X)$ into a unital $K^0(X)$-module.

One final connection between Ext and K-theory is that it can be used to solve the problem posed by Atiyah in [5]. Atiyah defined the set Ell (X) of generalized elliptic operators on a complex X and a map from Ell(X) onto the K-homology group $K_0(X)$ defined by Spanier-Whitehead duality. We show that an element of Ell(X) defines an element of $\mathrm{Ext}_0(X) \oplus Z$ yielding the diagram

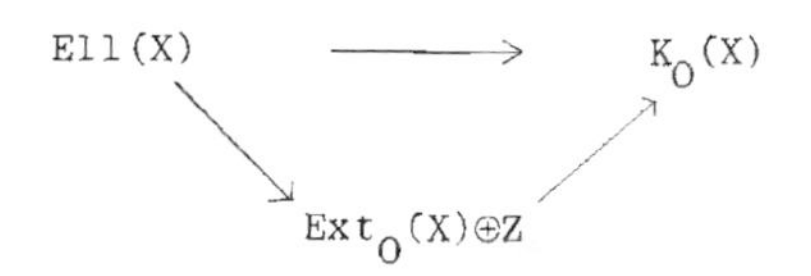

where the map from $Ext_0(X)\oplus Z$ to $K_0(X)$ is an isomorphism. A related solution to Atiyah's problem has been announced by Kasparov [12].

Extensions of $\mathcal{K}$ by the above C*-algebras belong to the type I theory. The beginnings of a corresponding theory for extensions in the type II situation has been made by Zsido [17] and Voiculescu and Zsido [16]. By using arguments similar to those already given, one can show that $Ext_R(X)$ is an abelian group and that $X \longrightarrow Ext_R(X)$ is a covariant functor. One expects this functor to define real K-homology, but certain technical details remain.

REFERENCES

1. M.B. Abrahamse, Toeplitz operators on multiply-connected regions, Amer. J. Math. 96 (1974), 261-297.
2. T.B. Andersen, Linear extensions, projections and split faces, J. Functional Anal. 17 (1974), 161-173.
3. T. Ando, Closed range theorems for convex sets and linear liftings, Pacific J. Math. 44 (1973), 393-410.
4. W.B. Arveson, A note on essentially normal operators, Proc. Roy. Irish Acad. Sect. A 74 (1974), 143-146.
5. M.F. Atiyah, Global theory of elliptic operators, Proc. Intern. Conf. on Functional Analysis and Related Topics (Toyko 1968), Univ. of Tokyo Press, Tokyo, 1970, pp. 21-30.
6. L.G. Brown, Characterizing Ext(X), this volume.
7. L.G. Brown, R.G. Douglas and P.A. Fillmore, Extensions of C*-algebras, operators with compact self-commutators, and K-homology, Bull. Amer. Math. Soc. 79 (1973), 973-978.
8. __________, Unitary equivalence modulo the compact operators and extensions of C*-algebras, Proc. of a Conf. on Operator Theory (Dalhousie 1973), Springer-Verlag Lecture Notes No. 345, Heidelberg, 1973, pp. 58-128.
9. __________, Extensions of C^*-algebras, and K-homology (submitted)
10. L.A. Coburn, Singular integral operators and Toeplitz operators on odd spheres, Indiana Univ. Math. J. 23 (1973), 433-439.

11. R.G. Douglas, Banach Algebra Techniques in the Theory of Toeplitz Operators, CBMS Regional Conference no. 15, Amer. Math. Soc., Providence, R.I., 1973.

12. G.G. Kasparov, The generalized index of elliptic operators, Functional Anal. Appl. 7 (1973), 82-83.

13. I.M. Singer, Prospects in Mathematics, No. 70, Annals Math. Studies, Princeton, 1971.

14. U. Venugopalkrishna, Fredholm operators associated with strongly pseudo convex domains in C^n, J. Functional Analysis 9 (1972), 349-373.

15. J. Vesterstrøm, Positive linear extensions operators for spaces of affine functions, Israel J. Math. 16 (1973), 203-211.

16. D. Voiculescu and L. Zsido, informal communication, 1974.

17. L. Zsido, The Weyl-von Neumann theorem in semi-infinite factors, J. Functional Analysis 18 (1975), 60-72.

BUNDLES AND SHEAVES ARE EQUIVALENT IN THE CATEGORY OF BANACH SPACES

Karl Heinrich Hofmann, Tulane University

Roughly speaking, a bundle is a function $p:E \to X$ of a topological space E, called bundle space, onto a topological space X, called base space; depending on the particular purpose at hand, various additional hypotheses are imposed, notably on the fibers $p^{-1}(x)$, $x \in X$ which are often assumed to be (topological) vector spaces whose operations behave well in relation to the induced topology. On the other hand, if $\mathcal{A}$ is a category, a pre-sheaf over a topological space X is a functor $\mathcal{F}:\mathcal{O}(X) \to \mathcal{A}$, where $\mathcal{O}(X)$ is the small category of open sets in X with an arrow $U \to V$ precisely when $V \subseteq U$. In order to speak of a sheaf we have to have (certain) limits in $\mathcal{A}$; a sheaf then is a pre-sheaf which preserves a particular type of limit which represents in functorial terms the idea of "patching local sections together". If the category $\mathcal{A}$ has direct (co-)limits one can associate with a presheaf a bundle $p:E \to X$ whose fibers (or stalks) are given by $p^{-1}(x) = \operatorname{colim}_U \mathcal{F}(U)$, where U ranges through the filter basis of open neighborhoods of x. If the category $\mathcal{A}$ is a category of discrete structures such as sets, abelian groups or vector spaces, one knows how to equip E with a topology such that all fibers inherit the discrete topology and p is a local homeomorphism. The space E is then called the sheaf space (espace étalé). If we have a bundle $p:E \to X$, such that for an open set $U \subseteq X$ the inverse image $p^{-1}(U)$ is in a category $\mathcal{A}$, then the rule $U \mapsto p^{-1}(U):\mathcal{O}(X) \to \mathcal{A}$ will define a presheaf in $\mathcal{A}$. In the case of a category with discrete objects one has very good control over the two processes connecting sheaves and bundles (or sheaf spaces). The situation is far less satisfactory in categories which are needed for applications in functional analysis.

Bundle techniques have an established history in the theory of operator algebras notably through the work of Gelfand, Godement, Fell,

Dixmier and Douady, Tomyama, Takesaki, to mention only a few authors of the earlier phase. A bibliography of more recent developments is available in my articles in Bull. Amer. Math. Soc. 78 (1972, 291-373 and Memoirs Amer. Math. Soc. 148 (1974), 177-182, and so no effort is made to give a complete system of references at present. Sheaf theoretic methods have been used for operator algebras in such contexts as the Arens-Calderon Theorem. In the algebraic structure theory of W*-algebras sheaves were extensively used by Teleman. An appropriate category for the discussion of bundle and sheaf techniques in functional analysis therefore is the category of Banach spaces. Since the very formulation of a sheaf requires that the category in question have products, the category of Banach spaces with <u>all</u> bounded operators is ruled out. We therefore consider the category <u>Ban</u> of all Banach spaces and non-expansive linear operators. This category is precisely suitable for the theory and, moreover, fully covers the applications to operator algebras.

<u>Our objective is to show that sheaf theory is possible in the category Ban and that it relates to an appropriately defined bundle theory as does sheaf theory for discrete abelian groups to the theory of sheaf spaces of abelian groups.</u>

In order to carry this program out one has to develop an explicit understanding for direct limits in <u>Ban</u>, and indeed for the later stages of the theory one has to operate in a category of Banach modules rather than Banach spaces. One is then prepared to introduce the concept of a pre-sheaf in the category <u>Ban</u>. For the introduction of the Banach space analog of the classical sheaf space we have to give a careful definition of a bundle of Banach spaces. The precise definition is new, but it turns out to be only a slightly generalized version of a bundle concept first formalized by Fell and then in a somewhat modified form by Dauns and myself. Equipped with this machinery we will be able to construct a presheaf from a bundle, and to associate with a bundle a pre-sheaf. We will then show that upon starting with a bundle, then passing to a presheaf and then to a bundle again we arrive at the point of departure. This is the easier portion of the theory. In order to accomplish the remainder, we will introduce the concept of a mono-presheaf in <u>Ban</u> and show that any such is isomorphically embedded into the presheaf of local sections of the associated bundle. Finally we define the concept of a sheaf. Any presheaf of sections of a bundle is in fact a sheaf. With the aid of an appropriate Stone-Weierstrass formalism we are able to show at least for hereditarily paracompact base spaces (and we note that all metrizable spaces are of this form!) that the sheaf of sections of the bundle associated with a given sheaf is isomorphic to the latter. The precise statement is complicated a bit by the fact that the given sheaf has to be a sheaf of $C(X)$-modules of a certain type. <u>Thus for a hereditarily paracompact base space X we obtain a natural bijection between certain sheaves of $C(X)$-modules and bundles over X.</u> We will illustrate the theory by outlining some concrete applications.

No proofs will be given in this presentation; the full details hopefully will appear elsewhere.

SECTION 1. THE CATEGORY OF BANACH SPACES AND MODULES

Throughout our discussion we fix the field of real or complex numbers as a field of scalars.

1.1 <u>DEFINITION</u>. Let A be a Banach algebra with identity. A <u>Banach module (over A)</u> or shortly an <u>A-module</u> is a Banach spaces M together with a bilinear map $(m,a) \mapsto ma : A \times M \to M$ which gives M the structure of a (unital) algebraic A-module and which satisfies

(i) $\|am\| \leq \|a\| \, \|m\|$.

This bilinear map is called the <u>scalar multiplication</u> of the module. A Banach subspace N of M is a <u>submodule</u> if $AN \subseteq N$.

1.2. <u>REMARK</u>. If N is a submodule of M , then M/N is an A-module relative to the scalar multiplication $(a, m+n) \to am+n$.

1.3 DEFINITION. Let A be a Banach algebra with identity. The category $\underline{Ban}_A$ of A-modules has as a class of objects the class of Banach modules over A and as class of morphisms the class of linear operators $f:M \to N$ between A-modules which are characterized by the following conditions:

(i) $f(am) = af(m)$ for all $a \in A$, $m \in M$.

(ii) $\|f(m)\| \le \|m\|$ for all $m \in M$.

If A is the ground field, we write Ban in place of $\underline{Ban}_A$ and call this category the category of Banach spaces.

One observes that the category $\underline{Ban}_A$ is complete and co-complete. The product in $\underline{Ban}_A$ of a family of A-modules M_j , $j \in J$, is the collection of all $(m_j)_{j \in J}$ in the cartesian product with $\sup_J \|m_j\| < \infty$; operations are componentwise, the norm is the sup-norm. For the purposes of sheaf theory we need an explicit description of the direct limit in the category $\underline{Ban}_A$. We assume familiarity with the direct limit in the category of vector spaces.

1.4 PROPOSITION. Let $\{E_j, j_{ji}, J\}$ be a direct system in $\underline{Ban}_A$ and let V be its direct limit in the category of vector spaces with $F_j:E_j \to V$ as colimit map. If $v \in V$, then $v = F_i(v_i)$ for some $i \in J$, $v_i \in E_i$ and the non-negative number

$$\|v\| = \lim_{i \le j} \|f_{ji}(v_i)\|$$

is well-defined (independently of j and v_j) and $v \mapsto \|v\|:V \to R^+$ is a seminorm. Let V_o be the zero-space of this seminorm and let E be the completion of the normed space V/V_o. Then E is the A-module and figures as the colimit of the system with colimit morphisms $f_j:E_j \to E$ derived from the F_j in the obvious way. Let $E_o = \cup\{f_j(E_j) : j \in J\}$. Then the following statements hold:

(i) E_o is dense in E .

(ii) For each $x \in E_o$ and each $x_i \in E_i$ with $f_i(x_i) = x$ one

<u>has</u> $\|x\| = \lim_{i \le j} \|f_{ji}(x_i)\|$.

By a short exact sequence in $\underline{Ban}_A$ we mean a sequence $0 \longrightarrow M_1 \xrightarrow{f} M_2 \xrightarrow{q} M_3 \longrightarrow 0$ in $\underline{Ban}_A$ such that f is an isometric embedding and q is quotient morphism. With this notation we have

1.5. <u>PROPOSITION.</u> <u>The direct limit functor preserves short exact sequences.</u>

Examples show that the direct limit functor fails to be exact in almost any other sense than that expressed in 1.5. Note that zero direct limits may arise from a sequence of isomorphisms of topological vector spaces: If E is any Banach space and $0 < r < 1$, then the sequence

$$E \xrightarrow{r} E \xrightarrow{r} E \xrightarrow{r} \ldots$$

has a zero direct limit in <u>Ban</u>.

SECTION 2. PRESHEAVES OF BANACH SPACES AND MODULES

If $(T,\le)$ is a partially ordered set, then T may be considered as a category with the elements of T as objects and an arrow $s \to t$ iff $t \le s$. The l.u.b. of a subset is then the limit in the sense of category theory. In this fashion, the set $\mathcal{O}(X)$ of open sets of a topological space X is made into a category. Thus we write $U \to V$ iff $V \subseteq U$; this may seem a bit unusual but appears to be convenient for our purposes.

2.1 <u>DEFINITION.</u> If X is a topological space, then a <u>presheaf</u> over X in $\underline{Ban}_A$ is a functor $\mathcal{J}:\mathcal{O}(X) \to \underline{Ban}_A$. In place of $\mathcal{J}(U \to V)$ we write $\mathcal{J}_{VU}$. The direct limit

$$\mathcal{J}(b) = \operatorname{colim}_{b \in U \in \mathcal{O}(B)} \mathcal{J}(U)$$

is called the <u>stalk</u> or the <u>fiber</u> of the presheaf $\mathcal{J}$ in b . We denote the colimit maps by $\mathcal{J}_{bU}:\mathcal{J}(U) \to \mathcal{J}(b)$.

2.2 NOTATION. Let $\mathcal{F}:\mathcal{O}(X) \to \underline{\mathrm{Ban}}_A$ be a presheaf. If $\sigma \in \mathcal{F}(U)$ and $V \subseteq U$ we write $\sigma|V = \mathcal{F}_{VU}(\sigma)$. If $b \in U$, we write $\hat{\sigma}(b) = \mathcal{F}_{bU}(\sigma)$. Thus $\hat{\sigma}$ is a function defined on U with $\hat{\sigma}(b) \in \mathcal{F}(b)$. We define

$$\begin{aligned}\mathcal{F}_o(b) &= \{x \in \mathcal{F}(b) : \text{there is an open neighborhood } U \text{ of } b \\ &\qquad \text{and an element } \sigma \in \mathcal{F}(U) \text{ such that } x = \hat{\sigma}(b)\} \\ &= \bigcup \{\mathcal{F}_{bU}(U) : b \in \mathcal{O}(X)\} .\end{aligned}$$

From the results of Section 1, notably of Proposition 1.4 then obtain

2.3 PROPOSITION. *For a presheaf $\mathcal{F}:\mathcal{O}(X) \to \mathrm{Ban}_A$ and $b \in X$ we have*

(i) *If* $U \in \mathcal{O}(X)$ *and* $\sigma \in \mathcal{F}(U)$, *then* $\|\hat{\sigma}\| \leq \|\sigma\|$, *where* $\|\hat{\sigma}\| = \sup_{u \in U}\|\hat{\sigma}(u)\|$.

(ii) $\mathcal{F}_o(b)$ *is dense in* $\mathcal{F}(b)$.

(iii) *For each* $x \in \mathcal{F}_o(b)$ *and each* $\sigma \in \mathcal{F}(U)$ *with* $x = \hat{\sigma}(b)$ *we have* $\|x\| = \lim_{b \in V \in \mathcal{O}(X), V \subseteq U}\|\sigma|V\|$.

We note the following consequence:

2.4 COROLLARY. *For* $\sigma \in \mathcal{F}(U)$ *the function* $b \mapsto \|\hat{\sigma}(b)\| : U \to \mathbb{R}^+$ *is upper semicontinuous.*

In the later parts of the theory, the so-called well-supported presheaves play an important role. In order to introduce them, we have to recall some notation.

2.4 DEFINITION. If M is an A-module and if $S \subseteq A$, $T \subseteq M$ we say that S and T *annihilate each other* iff $st = 0$ for all $(s,t) \in S \times T$, and we write $S \perp T$ when this relation holds. The set $\{a \in A : at = o \text{ for all } t \in T\}$ is called the annihilator of T (in A) written $T^{\perp}$. The annihilator of S (in M) is defined similarly.

Evidently M is a closed two-sided ideal of A. M is an $A/M^{\perp}$ module with the scalar multiplication given by $(a + M^{\perp})m = am$.

If I is a closed two-sided ideal of A and $p:A \to A/I$ is the quotient morphism, and if M is an A/I-module such that $p(a)m = am$ for all $a \in A$, $m \in M$, then $I \subseteq M^{\perp}$. In particular, if $\mathcal{J}:\mathcal{O}(X) \to \underline{Ban}_A$ is a presheaf, then every stalk $\mathcal{J}(b)$ is an $A/\mathcal{J}(b)^{\perp}$-module. If, in particular, $A = C(X)$ and if $I_b = \{f \in C(X) : f(b) = 0\}$, then the following two statements are equivalent:

(1) $fm = f(b)m$ for all $f \in C(X)$, $m \in M$

(2) $I_b \subseteq M^{\perp}$ (i.e. $I_b \perp M$).

This gives the easy part of the following proposition:

2.6 PROPOSITION. *Let $\mathcal{J}:\mathcal{O}(X) \to Ban_{C(X)}$ be a presheaf. Then the following statements are equivalent:*

(1) $(f\sigma)^{\wedge}(b) = f(u)\hat{\sigma}(u)$ *for all* $\sigma \in \mathcal{J}(U)$, $u \in U$.

(2) $I_b \perp \mathcal{J}(b)$ *for all* $b \in X$.

These conditions are implied by

(3) $I_U \perp \mathcal{J}(U)$ *for all* $U \in \mathcal{O}(X)$.

If the natural morphism $\sigma| \to \hat{\sigma} : \mathcal{J}(U) \to \Pi_{u\in U}\mathcal{J}(u)$ *is always injective, then all three conditions are equivalent.*

We are now ready for the following definitions:

2.7 DEFINITION. A presheaf $\mathcal{J}:\mathcal{O}(X) \to \underline{Ban}_{C(X)}$ is called *well-supported*, if it satisfies condition (3) of 2.6.

SECTION 3. BUNDLES OF BANACH SPACES

If $p_k:E_k \to X$, $k = 1,2$ are two (continuous) functions we denote with $E_1 \times_X E_2$ the fibered product (or pullback) over X given by $\{(x,y) \in E_1 \times E_2 : p_1(x) = p_2(y)\}$; there is an evident map $p:E_1 \times_X E_2 \to X$ given by $p(x,y) = p_1(x)$.

3.1 DEFINITION. a) A *family of Banach spaces* is a surjective function $p:E \to X$ such that $E_b = p^{-1}(b)$ is a Banach space for all $b \in X$.

b) For any family $p:E \to X$ of Banach spaces we can define the following functions:

(1) an <u>addition</u> add: $E \times_X E \to$, add$(x,y) = x + y$ in $E_{p(x)}$,

(2) a <u>scalar multiplication</u> scal: $K \times E \to E$, scal$(r,x) = rx$ in $E_{p(x)}$ (where $K = R,C$) ,

(3) a <u>zero selection</u> $O: X \to E$, $O(b)$ = zero of E_b ,

(4) a norm $\| \ \| : E \to R^+$, $\|x\|$ = norm of x in $E_{p(x)}$.

c) A <u>selection</u> for a family $p:E \to X$ of Banach spaces is a function $\sigma:Y \to E$, $Y \subseteq X$ with $p\sigma(y) = y$ for all $y \in Y$. If $Y = X$, then σ is a <u>global selection</u>. A selection σ is <u>bounded</u> if $\{\|\sigma(y)\| : y \in Y\}$ is a bounded set.

If $\sigma : Y \to E$ is a selection and $r > 0$, then w write $U(\sigma,r) = \{x \in E : p(x) \in Y$ and $\|x - \sigma(p(x))\| < r\}$, and call this set the <u>tube of radius r around</u> σ .

3.2 <u>DEFINITION.</u> A <u>prebundle of Banach spaces</u> (or shortly a prebundle) is a family of Banach spaces $p : E \to X$ satisfying the following axioms:

AXIOM I. The set E carries a topology such that

(0) The induced topology and the Banach space topology agree on E_b for all $b \in X$.

(1) Addition add: $E \times_X E \to E$ is continuous.

(2) Scalar multiplication scal: $K \times E \to E$ is continuous.

(3) The norm $\| \ \| : E \to R^+$ is upper semicontinuous, i.e. all tubes $U(O,r)$ are open.

In the presence of (2), condition (3) is equivalent to

(4) The tube $U(O,1)$ is open.

AXIOM II. The set X carries a topology such that

(1) $p : E \to X$ is continuous and open.

(2) The zero selection $O : X \to E$ is continuous.

AXIOM III. For each $b \in X$ the collection of all $U(O|V,r)$, V an open neighborhood of B , and $r > 0$ is a neighborhood basis of $O(b)$.

The space E is called the *fiber space*, X is the *base space*, and the E_b are the *fibers*. A selection is a *section* if it is continuous.

3.3 *DEFINITION.* A *bundle of Banach spaces* (or shortly a *bundle*) is a prebundle $p : E \to X$ satisfying the following axiom:

AXIOM IV. For each $x \in E$ and each $\epsilon > 0$ there is a section $\sigma : U \to E$ with $p(x) \in U$ and $\|x - \sigma(p(x))\| < \epsilon$.

A *full bundle* is a bundle satisfying the stronger axiom

AXIOM V. For each $x \in E$ there is a global section with $x - \sigma(p(x))$.

Clearly both axioms IV and V secure an abundance of sections; Axiom IV has not been considered before. The following result due to Douady and Dal Soglio-Herault shows that it is not easy to come by examples of bundles which are not full. We will see however, that bundles naturally arise in the context of sheaves.

3.4 *PROPOSITION.* *Every prebundle of a locally paracompact base space is a full bundle.*

Here a space is called *locally paracompact* if every point has at least one closed (!) neighborhood which is paracompact. Each such space is automatically completely regular. The following is a parallel result based on a result of Dupré's:

3.5 *PROPOSITION.* *Every bundle over a uniformizable space is full.*

We now proceed to the construction of prebundles and bundles.

3.6 *PROPOSITION.* *Let* $p : E \to X$ *be a family of Banach spaces. Suppose that the following data are given:*

(a) *a topology on* X ,

(b) *a set* S *of selections* $s: V \to E$, $V \in \mathcal{O}(X)$, *and suppose that the following conditions are satisfied:*

i) *All functions* $v \mapsto \|s(v)\| : V \to R^+$, $s \in S$ *are upper semicontinuous.*

ii) *For all* $b \in X$ *the set* $E_b \cap \bigcup \{\text{im } s: s \in S\}$ *is dense in* E_b .

iii) *If* $s: V \to E$ *is in* S *and* $W \in \mathcal{O}(X)$, *then* $s|W \in S$. *In particular, the empty selection is in* S . *If* m $s: V \to E$ *and* $t: W \to W$ *are in* S , *then* $s + t = s|(V \cap W + t|(V \cap W)$ *is in* S . *If* $s \in S$, *then* $ks \in S$ *for any scalar* k .

Conclusion: *There is a unique* coarsest topology on E such that $p: E \to X$ *is a bundle and all* $s \in S$ *are local sections. If in place of* ii) *we have the stronger hypothesis*

ii') $E = \bigcup\{\text{im } s : s \in S\}$,

then there is only one such topology on E .

The proof of this proposition is rather long and technical; however, it is at the root of virtually every existence theorem for bundles arising in functional analysis. It shows, in particular that in a bundle the topology of the fiber space is uniquely determined by the other data. The following corollary is frequently applied:

3.7 *COROLLARY.* *Let* $p: E \to X$ *be a family of Banach spaces and suppose that a topology* $\mathcal{O}(X)$ *is given. If* S *is a vector space of bounded global selections such that* $S(X) = E$ *and that all* $b \mapsto \|s(b)\| : X \to R^+$, $s \in S$ *are upper semicontinuous, then there is a unique topology on* E *making* $p: E \to X$ *into a full bundle such that all* $s \in S$ *are continuous.*

We will now discuss how presheaves arise from prebundles and vice versa.

Firstly let $p:E \to X$ be a prebundle. For each open set $U \subseteq X$ we let $\Gamma(U)$ be the set of all bounded sections defined on U. One observes that $\Gamma(U)$ is a Banach space under pointwise operations and with the sup-norm; in fact $\Gamma(U)$ is a $C(X)$-modules under the scalar multiplication $(f,\sigma) \mapsto f\sigma$ which is defined by $(f\sigma)(u) = f(u)\sigma(u)$ for all $u \in$ domain σ. If $V \subseteq U$ in $\mathcal{O}(X)$ we define $\Gamma_{VU} : \Gamma(U) \to \Gamma(V)$ by $\Gamma_{VU}(\sigma) = \sigma|V$. We then have the following result:

3.8 PROPOSITION. *Let* $p:E \to X$ *be a prebundle. Then* $\Gamma:\mathcal{O}(X) \to \underline{\text{Ban}}_{C(X)}$ *is a well-supported presheaf of* $C(X)$*-modules over* X.

The presheaf Γ is called the presheaf associated with p.

Secondly we consider a presheaf $\mathcal{J}:\mathcal{O}(X) \to \underline{\text{Ban}}$ of Banach spaces. We let $E = \bigcup_{b \in X} \mathcal{J}(b) \times \{b\}$ and define $p:E \to X$ by $p(x,b) = b$. We take the set S of all functions $u \mapsto (\hat{\sigma}(u),u):U \to E$ with $\sigma \in \mathcal{J}(U)$, $U \in \mathcal{O}(X)$. One verifies that the hypotheses of 3.5 are satisfied and obtains

3.9 PROPOSITION. *For any presheaf* $\mathcal{J}:\mathcal{O}(X) \to \underline{\text{Ban}}$ *there is a unique coarsest topology on* $E = \bigcup_{b \in X}\mathcal{J}(b) \times \{b\}$ *making* $p:E \to X$ *into a bundle such that all selections* $u \mapsto (\hat{\sigma}(u),u):U \to E$, $\sigma \in \mathcal{J}(U)$, $U \in \mathcal{O}(X)$ *are continuous.*

The bundle $p:E \to X$ is called the *bundle associated with the presheaf* $\mathcal{J}$.

We now have natural processes to obtain a presheaf from a bundle and vice versa. What happens if we iterate these procedures? This question is relatively easy to handle if we start from a bundle and return to a bundle. The other case is more difficult.

3.10 PROPOSITION. *Let* $p:E \to X$ *be a prebundle and* $\Gamma:\mathcal{O}(X) \to \underline{\text{Ban}}_{C(X)}$ *the associated well-supported presheaf of* $C(X)$*-modules. Let* $p':E' \to X$ *be the bundles associated with* Γ. *Then there is a*

commutative diagram

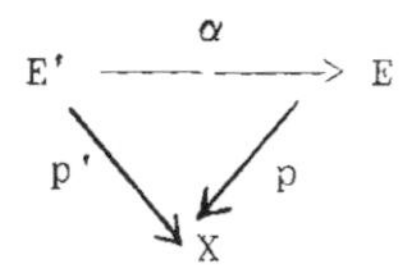

in which α *is a bijective open map onto its image and induces an isometric embedding on each stalk. Also,* $\alpha(E'_b)$ *is the smallest closed subvector space of* E_b *containing all elements which lie on some local section of* E .

There are some questions left open which are due to the fact that we do not know for sure whether the topology of the fiber space of a bundle is generally determined by the other data. For full bundles this ambiguity vanishes, and we have the following theorem:

3.11 THEOREM. *Let* $p:E \to X$ *be a full bundle and* $\Gamma:\mathcal{O}(X) \to \underline{\mathrm{Ban}}_{C(X)}$ *the associated presheaf. Let* $p':E \to X$ *be the bundle associated with* Γ . *Then* p *and* p' *are isometrically isomorphic.*

If p fails to be full, then in general we can only conclude that p' is isomorphic to the unique coarsification of p (which exists as a consequence of 3.6). Recall that p is automatically full if X is uniformizable.

We now begin to investigate the alternative iteration: We begin with a presheaf $\mathcal{J}:\mathcal{O}(X) \to \underline{\mathrm{Ban}}$ and associate with it the canonical bundle $p:E \to X$. Now let $\Gamma:\mathcal{O}(X) \to \underline{\mathrm{Ban}}_{C(X)}$ be the presheaf associated with p . For each $U \in \mathcal{O}(X)$ there is a $\underline{\mathrm{Ban}}$-morphism $\gamma_U:\mathcal{J}(U) \to \Gamma(U)$ given by $\gamma_U(\sigma)(u) = (\hat{\sigma}(u),u)$. One notes that $\gamma:\mathcal{J} \to \Gamma$ is a natural transformation of functors. If $\mathcal{J}$ is in fact a well-supported presheaf of $C(X)$-modules, then all γ_U are $C(X)$-module maps, and γ is a morphism of presheaves of $C(X)$-modules. In order to investigate the morphism γ more carefully, we need some additional concepts. The first is that of a mono-presheaf. Let

$\mathcal{F}:\mathcal{O}(X) \to$ Ban be a presheaf. For a given $U \in \mathcal{O}(X)$ we consider an open cover $\mathcal{C} = \{V_j : j \in J\}$ of U and define a morphism in Ban

$$\phi_{\mathcal{C}} : \mathcal{F}(U) \to \Pi_{j\in J}\, \mathcal{F}(V_j) \ , \ \phi_{\mathcal{C}}(\sigma) = (\sigma|V_j)_{j\in J} \ .$$

we then derive

3.12 PROPOSITION. *Let $\mathcal{F}:\mathcal{O}(X) \to$ Ban be a presheaf and let $U \in \mathcal{O}(X)$. Then the following conditions are equivalent.*

(1) *For each $\sigma \in \mathcal{F}(U)$, each open cover $\{V_j : j \in J\}$, and each $\epsilon > 0$ there is a $k \in J$ such that $\|\sigma|V_k\| > \|\sigma\| - \epsilon$.*

(2) *For each open cover $\mathcal{C} = \{V_j : j \in J\}$ of U, the morphism $\phi_{\mathcal{C}}: \mathcal{F}(U) \to \Pi_{j\in J}\mathcal{F}(V_j)$ is an isometric embedding.*

(3) *$\gamma_U : \mathcal{F}(U) \to \Gamma(U)$ is an isometric embedding.*

3.13 DEFINITION. A presheaf $\mathcal{F}:\mathcal{O}(X) \to$ Ban satisfying the equivalent conditions of 3.12 for all $U \in \mathcal{O}(X)$ is called a *mono-presheaf*.

One notes that a mono-presheaf $\mathcal{F}:\mathcal{O}(X) \to \mathrm{Ban}_{C(X)}$ of $C(X)$-modules is well-supported iff $\gamma:\mathcal{F} \to \Gamma$ is a $C(X)$-module map.

In order to complete the theory we finally have to introduce particular mono-presheaves; we do this in the following section.

SECTION 4. SHEAVES OF BANACH SPACES AND MODULES

4.1 DEFINITION. A *sheaf of Banach modules* is a presheaf $\mathcal{F}:\mathcal{O}(X) \to \mathrm{Ban}_A$ satisfying the following additional conditions:

For any family $\{V_j : j \in J\}$ of open sets in X and each family $\{\sigma_j : j \in J\}$ elements $\sigma_j \in \mathcal{F}(V_j)$ satisfying $\sigma_j|V_j \cap V_k = \sigma_k|V_j \cap V_k$ for all $(j,k) \in J \times J$, there is one and only one $\sigma \in \mathcal{F}(U)$, $U = \bigcup_{j\in J} V_j$ such that $\sigma|V_j = \sigma_j$ for all $j \in J$.

This is the familiar definition of a sheaf in a complete category and may be expressed in terms of products and equalizers or, alternatively, as the preservation of certain limits. Every sheaf is a mono-presheaf, while the converse fails.

The presheaf associated with a prebundle is always a sheaf in general. With the aid of the sheaf condition we show that for every sheaf $\mathcal{J}:\mathcal{O}(X) \to \underline{\text{Ban}}$ and each $U \in \mathcal{O}(X)$, the vector space $\gamma_U(\mathcal{J}(U))$ is fully additive in $\Gamma(U)$. Here we say that a subvector space S of the vector space $\Pi_{b\in X}E_b$ (for a family of Banach spaces $p:E \to X$) is _fully additive_ if every locally finite family $\{\sigma_j : j \in J\}$ of elements of S whose sum is bounded has its sum in S. Moreover, if $\mathcal{J}:\mathcal{O}(X) \to \underline{\text{Ban}}_{C(X)}$ is a well-supported sheaf over a regular space X, then for any uniformizable open subset U of X and any $u \in U$ the colimit map $\mathcal{J}_{uX}:\mathcal{J}(X) \to \mathcal{J}(u)$ has dense image. This remark togeth - her with an appropriate Stone-Weierstrass argument enters into the proof of the following result:

4.2 PROPOSITION. _Let $\mathcal{J}:\mathcal{O}(X) \to \underline{\text{Ban}}_{C(X)}$ be a well-supported sheaf over a regular space. If U is a paracompact open subset of S, then $\gamma_U:\mathcal{J}(U) \to \Gamma(U)$ is an isometric isomorphism. Moreover, for each $u \in U$, the colimit map $\mathcal{J}_{uU}:\mathcal{J}(U) \to \mathcal{J}(u)$ is a quotient morphism._

This allows us to settle the question for the relation between sheaves and bundles at least for a large class of base spaces. Recall that a space is _hereditarily paracompact_ if every open subset is again paracompact. This is in fact equivalent to the condition that every subspace be again paracompact. All metrizable spaces have this property. The following theorem then is a counterpart to Theorem 3.11

4.3 THEOREM. _Let $\mathcal{J}:\mathcal{O}(X) \to \underline{\text{Ban}}$ be a sheaf over a hereditarily para-compact space X. If $p:E \to X$ is the bundle associated with $\mathcal{J}$ and $\Gamma:\mathcal{O}(X) \to \underline{\text{Ban}}_{C(X)}$ the well-supported sheaf of $C(X)$-modules asso-ciated with p, then the following conditions are equivalent:_

(1) _$\mathcal{J}$ is a well-supported sheaf of $C(X)$-modules._

(2) _$\gamma:\mathcal{J} \to \Gamma$ is an isometric isomorphism of sheaves._

We can now formulate the fundamental theorem for hereditarily paracompact base spaces. All of these spaces are locally paracompact, hence in particular completely regular. Thus all bundles over such spaces are full.

4.4 FUNDAMENTAL THEOREM ON SHEAVES AND BUNDLES OF BANACH SPACES

<u>Let X be a hereditarily paracompact space. Then there is a natural bijection between the isometric isomorphy classes of bundles of Banach spaces over X and the isometric isomorphy classes of well-supported sheaves of $C(X)$-modules over X. The bijection is implemented by associating with a bundle its sheaf of bounded sections and with a sheaf its associated bundle.</u>

SECTION 5. SOME EXAMPLES

We sample a few natural occurences of bundles and presheaves in functional analysis.

5.1 <u>EXAMPLE.</u> Let A be a unital Banach algebra and $X = \text{Prim } A$ its primitive ideal space with the hull-kernel topology. For $Y \subseteq X$ let $k(Y) = \bigcap \{I: I \in Y\}$. If $Z \subseteq Y$ then $k(Y) \subseteq k(Z)$, so there is a canonical morphism $A/k(Y) \to A/k(Z)$. Thus we obtain a presheaf of Banach algebras

$$\mathcal{J}: \mathcal{O}(X) \to \underline{\text{Banalg}}, \quad \mathcal{J}(U) = A/k(U).$$

From the exact sequences

$$0 \longrightarrow k(U) \longrightarrow A \longrightarrow \mathcal{J}(U) \longrightarrow 0$$

we obtain for each $I \in X$ an ecact seauence

$$0 \longrightarrow \widetilde{I} \longrightarrow A \longrightarrow \mathcal{J}(I) \longrightarrow 0$$

with $\mathcal{J}(I) \cong A/\widetilde{I}$, where $\widetilde{I} = (\bigcup\{k(U): I \in U \in \mathcal{O}(X)\})^{-}$. We have in fact

$$k(U) = \bigcap \{\widetilde{I}: I \in U\} \quad \text{for all } U \in \mathcal{O}(X).$$

This allows us to conclude that $\mathcal{J}$ is a mono-presheaf. For C^*-algebras we may rewrite

$\tilde{I} = (\bigcup\{a^\perp : a \notin I\})^-$, $a^\perp$ = largest closed ideal annihilating a .

In general, $\mathcal{J}$ is not a sheaf; however, for each open closed subset U of X the sheaf condition of 4.1 is satisfied. Furthermore, one can show that $\mathcal{J}$ is a well-supported presheaf of $C(X)$-modules; this is the so-called Dauns-Hofmann theorem which says that each unital C^*-algebra A is a $C(X)$-module in such a fashion that $fa \in f(I)a + I$ for all $I \in X$.

Summarizing we have the following result:

Let A be a C^*-algebra with identity. Then the presheaf $\mathcal{J}$ associated with A is a well-supported mono-presheaf $\mathcal{J}:\mathcal{O}(X) \to \text{Ban}_{C(X)}$ $X = \text{Prim } A$. For every compact open set $U \in \mathcal{O}(X)$ the map $\gamma_U:\mathcal{J}(U) \to \Gamma(U)$ is an isomorphism. If U and $\bar{U}$ are open, then $\mathcal{J}_{U\bar{U}}:\mathcal{J}(\bar{U}) \to \mathcal{J}(U)$ is the identity. If X is Hausdorff then $A \cong \Gamma(X)$.

5.2 EXAMPLE. Let M be a $C(X)$ module. For any ideal J of $C(X)$ let JM be the closed submodule generated by all fm , $f \in J$, $m \in M$. For $Y \subseteq X$ we write $I_y = \{f \in C(X) : f(y) = 0 \text{ for } y \in Y\}$. We now define $\mathcal{J}:\mathcal{O}(X) \to \text{Ban}_{C(X)}$ by the exact sequence

$$0 \longrightarrow I_U M \longrightarrow M \longrightarrow \mathcal{J}(U) \longrightarrow 0 .$$

Then $\mathcal{J}$ is a well-supported presheaf. The exact colimit sequence is

$$0 \longrightarrow I_b M \longrightarrow M \longrightarrow \mathcal{J}(b) \longrightarrow 0 ,$$

whence the fiber $\mathcal{J}(b)$ is $M/I_b M$. By a result of Varela we have $\|m + I_Y M\| = \inf\{\|fm\| : f \in F_Y\}$ where $F_Y = \{f \in C(X) : f(X) \subseteq [0,1]$, $f^{-1}(1)$ is a neighborhood of $\bar{Y}$ in $X\}$.

We say that M is locally $C(X)$-convex if $f,g > 0$ in $C(X)$ and $m,n \in M$ implies $\|fm + gn\| \le \|f + g\| \max\{\|m\| , \|n\|\}$. If $p:E \to X$ is a prebundle, then every closed $C(X)$ submodule of $\Gamma(X)$ is a locally $C(X)$-convex $C(X)$-module. We have the following result:

Let M be a locally $C(X)$-convex $C(X)$-module for a compact space

X . <u>If $\mathcal{J}:\mathcal{O}(X) \to \underline{Ban}_{C(X)}$ be the presheaf associated with M and if $U \in \mathcal{O}(X)$ is open closed, then $\gamma_U:\mathcal{J}(U) \to \Gamma(U)$ is an isometric isomorphism.</u> (<u>This applies in particular to</u> $U = X$.) <u>If U and $\bar{U}$ are open, then $\mathcal{J}_{U\bar{U}}:\mathcal{J}(\bar{U}) \to \mathcal{J}(U)$ is the identity.</u>

This gives us the following structure theorem for $C(X)$-modules:

<u>Let M be a $C(X)$-module for a compact space X and $p:E \to X$ the bundle associated to the presheaf of M . The fibers are $E_b \cong M/I_bM$. Then the following statements are equivalent:</u>

(1) M <u>is locally $C(X)$-convex.</u>

(2) $\gamma_X:M \to \Gamma(X)$ <u>is an isometric isomorphism of $C(X)$-modules.</u>

(3) M <u>is isometrically isomorphic to some $C(X)$-module of global sections of some bundle over</u> X .

As Varela has observed, for any compact space X the Banach space $M(X)$ of all Radon measures is a $C(X)$-module (under multiplication) such that $\gamma_X:M(X) \to \Gamma(X)$ annihilates all continuous measure Thus $M(X)$ is not locally $C(X)$-convex, if X allows continuous measures.

We illustrate the previous results by taking a C^*-algebra A and let Z be the centroid (i.e. the center of the multiplier algebra). Then $Z = C(\text{Max } Z)$. One notes that A is a locally Z-convex Z-module and obtains the following result (Dauns, Hofmann, Varela):

<u>Any C^*-algebra is isomorphic to the C^*-algebra of global sections in a bundle $p:E \to \text{Max } Z$ of C^*-algebras, and E_I is isomorphic to</u> A/IA , $I \in \text{Max } Z$.

TOPOLOGICAL OBSTRUCTIONS TO PERTURBATIONS OF PAIRS OF OPERATORS

Jerome Kaminker, Indiana-Purdue Univ. and Claude Schochet, Indiana[1]

1. Introduction; the main theorem.

Given a compact metric space X, L.G. Brown, R. G. Douglas and P. A.Fillmore [1 - 4] (referred to as BDF) have defined an abelian group Ext(X). When $X \subseteq R^2$ these groups measure the obstruction to compactly perturbing an essentially normal operator to a normal operator. Various other operator-theoretic results are related to Ext(X) for X of low dimension (c.f. BDF [1], Brown-Schochet [5].) BDF [1] have characterized Ext(X) for $X \subseteq R^3$. In the present paper we characterize Ext(X) for $X \subseteq R^4$ and show how this characterization yields operator-theoretic information. Our main theorem relates Ext to homology:

Main Theorem. Let X be a compact subset of R^4. Then there is an isomorphism of abelian groups

$$\mathrm{Ext}(X) \cong {}^s\tilde{H}_1(X) \oplus {}^s\tilde{H}_3(X) \tag{1}$$

where ${}^s\tilde{H}_j$ is (reduced) Steenrod homology theory. The isomorphism is natural on finite complexes (in which case ${}^s\tilde{H}_n = \tilde{H}_n$, singular homology).

Steenrod homology was introduced by Steenrod [10] and axiomatized by Milnor [9]. We prefer to use it since singular homology is not well-behaved on compact metric spaces. Further, it bears a strong relationship to Ext, as will be explained below.

The following short exact sequences indicate the relationship of ${}^s\tilde{H}_*$ to Čech cohomology and Cech homology. Recall, if $X = \varprojlim X_j$, each X_j a finite complex, then $\check{H}_*(X) = \varprojlim \tilde{H}_*(X_j)$ and $\check{H}^*(X) = \varinjlim \tilde{H}^*(X_j)$.

Universal Coefficient Theorem (Eilenberg-MacLane [6])

$$0 \to \mathrm{Ext}^1(\check{\tilde{H}}^{n+1}(X), Z) \to {}^s\tilde{H}_n(X) \to \hom(\check{\tilde{H}}^n(X), Z) \to 0 \tag{2}$$

Lim^1 -sequence: (Steenrod [10]).

$$0 \to \varprojlim{}^1 \tilde{H}_{n+1}(X_j) \to {}^s\tilde{H}_n(X) \to \check{\tilde{H}}_n(X) \to 0 \tag{3}$$

---

[1] Research partially supported by a grant from the National Science Foundation.

(The functor $\varprojlim^1$ is defined below.). We now indicate the application of the main theorem to operator theory.

2. The two-operator perturbation problem.

Let $\mathcal{L}$ be the bounded operators on a complex, infinite-dimensional, separable Hilbert space, with $\mathcal{K}$ the ideal of compact operators and $\pi: \mathcal{L} \to \mathcal{L}/\mathcal{K}$. Suppose $A_1, A_2 \in \mathcal{L}$ and suppose that $\pi A_1, \pi A_2$ are commuting normals. When do there exist commuting normal operators $B_1, B_2 \in \mathcal{L}$ with $A_i - B_i \in \mathcal{K}$? BDF tells us that a necessary condition for this perturbation to exist is

$$\text{index } (A_i - \lambda I) = 0 \text{ where defined,} \tag{4}$$

but even if (4) is satisfied the operators produced may not commute. The joint perturbation problem does have an expression in terms of Ext, as was observed by BDF (the case of two operators is stated here, but the n-operator version makes good sense as well).

Let $X_i = \sigma_e(A_i)$ and $X = \text{joint } \sigma_e(A_1, A_2)$.
Let $\tau = \tau[A_1, A_2] \in \text{Ext}(X)$ be the equivalence class of the extension

$$0 \to \mathcal{K} \to C^*\{I, \mathcal{K}, A_1, A_2\} \to C(X) \to 0.$$

It is almost immediate that the perturbation exists if and only if $\tau = 0$, the identity in Ext(X). Note that $X \subseteq R^4$, so the main theorem (1) applies.

3. Analysis of the topological obstructions.

Recall that the main theorem (1) provides an isomorphism

$$\text{Ext}(X) = {}^s\tilde{H}_1(X) \oplus {}^s\tilde{H}_3(X)$$

and hence projections $\Gamma_j: \text{Ext}(X) \to {}^s\tilde{H}_j(X)$, $j = 1, 3$. Let $\tau \in \text{Ext}(X)$ be the class defined in §2. Then $\tau = 0$ if and only if $\Gamma_3\tau = \Gamma_1\tau = 0$. Let us examine these maps separately.

The group ${}^s\tilde{H}_3(X)$ is isomorphic to $\check{\tilde{H}}^0(R^4 \backslash X)$, the group of locally constant, integer-valued functions on the bounded components of the complement of X in R^4, by Steenrod duality [9]. So $\Gamma_3\tau$ is a function which assigns an integer to each bounded component of the complement of X. It would seem to be a "higher" index map. The map Γ_3 is natural for all $X \subseteq R^4$. Note that if X does not separate R^4 then $\Gamma_3 = 0$.

The map Γ_1 is related to the BDF index map

$$\gamma : \mathrm{Ext}(Y) \to \hom(\pi^1(Y), Z)$$

as follows. Let ρ_i be the composite $X \to X_1 \times X_2 \to X_i$. Then for each i there is a natural commutative diagram

$$\begin{array}{ccccc} \mathrm{Ext}(X) & \xrightarrow{\rho_{i*}} & \mathrm{Ext}(X_i) & \xrightarrow{\gamma} & \hom(\pi^1(X_i), Z) \\ \downarrow \Gamma_1 & & \downarrow \Gamma_1 & & \downarrow D \\ {}^s\tilde{H}_1(X) & \xrightarrow[\rho_{i*}]{} & {}^s\tilde{H}_1(X_i) & \xrightarrow[\omega]{} & \hom(\tilde{H}_0(C \setminus X_i), Z) \end{array} \tag{5}$$

where D is induced by Alexander duality and ω makes the diagram commute. Hence, if K is a bounded path component of $C \setminus X_i$, (that is, a basis element of the free abelian group $\tilde{H}_0(C \setminus X_i)$), one has

$$\begin{aligned} \omega\rho_{i*}\, \Gamma_1(\tau)(K) &= D\gamma\rho_i^*(\tau)(K) \\ &= D\gamma[A_i](K) \\ &= \text{index } (A_i - \lambda I) \qquad \text{for any } \lambda \in K \end{aligned}$$

Thus Γ_1 contains the BDF index information (4) which was previously observed to be necessary.

However, the map Γ_1 is considerably more complex, on several counts. First, the maps ρ_{i*} need not be mono. For example, take $S = A_1 + iA_2$ to be the unilateral shift, written as the sum of its real and imaginary parts. Then $\sigma_e A_1 = \sigma_e A_2 = [0,1]$ but joint $\sigma_e(A_1, A_2) = \sigma_e(S) = S^1$. So applying the BDF criterion (4) separately to the A_i yields no information, yet the perturbation is impossible in that case. (If it were possible then S would be of the form normal + compact, and Fredholm index information prohibits that.)

Second, the group $\mathrm{Ext}(\tilde{K}^0)X), Z)$ (which in fact sits as a natural subgroup of Ext(X)) is annihilated by the index maps γ and γ_∞. For example, take X to be the real projective plane. Then

$\mathrm{Ext}(X) = Z/2Z$, ${}^{s}\tilde{H}_1(X) = Z/2Z$, ${}^{s}\tilde{H}_3(X) = 0$, Γ_1 is an isomorphism, $\Gamma_3 = 0$, but the index invariants vanish. More perversely, let X be the suspension of the 2-adic solenoid. Again Γ_1 is an isomorphism, but $\mathrm{Ext}(X) = {}^{s}\tilde{H}_1(X) = \mathrm{Ext}(\check{H}^2(X),Z) = \hat{Z}_2/Z$, where $\hat{Z}_2$ is the group of 2-adic integers!

To sum up, then, the fact that $X \subseteq R^4$ (rather than R^2) allows the term ${}^{s}\tilde{H}_1(X)$ to be considerably more complicated. Condition (4) of BDF is a first-line approximation to the invariant Γ_1, but only that.

We now move towards a proof of the main theorem. This requires a digression to Steenrod homology.

4. Generalized Steenrod homology

Definition. [7] A generalized Steenrod homology theory is a sequence of covariant, homotopy-invariant functors from compact metric spaces to abelian groups satisfying the following three axioms for all n:

Exactness Axiom: If A is a closed subset of X, then the sequence

$$h_n(A) \to h_n(X) \to h_n(X/A)$$

is exact.

Suspension Axiom. There is a natural isomorphism $h_n(X) \simeq h_{n+1}(SX)$. (where SX is the (unreduced) suspension of X).

Wedge Axiom. If $X = \vee_j X_j$ is the strong wedge of a countable family of pointed compact metric spaces, then

$$h_n(X) \simeq \Pi_j h_n(X_j)$$

(where $\vee_j X_j$ is the subspace of $\Pi_j X_j$ consisting of those points which differ from the basepoint in at most one coordinate.) Here is the strong wedge of a countable family of circles:

There are two relevant examples. The first is Steenrod homology

${}^s\tilde{H}_*$. It also satisfies the dimension axiom: ${}^s\tilde{H}_j(S^0) = 0$ for $j \neq 0$ and ${}^s\tilde{H}_0(S^0) = Z$. Milnor [9] shows that the above axioms uniquely characterize Steenrod homology.

The second example is the homology theory E_* obtained from Ext by defining

$$E_n(X) = \begin{cases} \mathrm{Ext}(X) & n \text{ odd} \\ \mathrm{Ext}(SX) & n \text{ even} \end{cases}$$

This depends critically upon the BDF generalization of Bott periodicity to compact metric spaces: $\mathrm{Ext}(S(SX)) \simeq \mathrm{Ext}(X)$, (c.f. BDF [3]).

The axioms imply that Milnor's (homology) $\varprojlim^1$-theorem holds. If

$$G_0 \xleftarrow{\alpha_1} G_1 \xleftarrow{\alpha_2} G_2 \xleftarrow{\alpha_3} \cdots$$

is an inverse sequence of abelian groups and homomorphisms, let

$\psi: \Pi_j G_j \to \Pi_j G_j$ by $\psi(g_0, g_1, g_2, \ldots) = (g_0 - \alpha_1 g_1, g_1 - \alpha_2 g_2, g_2 - \alpha_3 g, \ldots)$.

Then define

$$\varprojlim G_j = \text{kernel } (\psi)$$

and

$$\varprojlim^1 G_j = \text{cokernel } (\psi) = \Pi_j G_j / \text{image}(\psi)$$

If $X = \varprojlim X_j$ is the inverse limit of a sequence of finite complexes, then define

$$h_n(X) = \varprojlim h_n(X_j)$$

and

$$\mathcal{L}h_n(X) = \varprojlim^1 h_{n+1}(X_j).$$

Milnor's $\varprojlim^1$-sequence for E_* then reads

$$0 \to \mathcal{L}E_n(X) \to E_n(X) \to \check{E}_n(X) \to 0. \qquad (6)$$

The group $\mathcal{L}E_n(X)$ measures the lack of truth in the statement $\varprojlim E_n = E_n \varprojlim$. Note that $\mathcal{L}E_n(X)$ is divisible, hence (6) yields an unnatural isomorphism

$$E_n(X) \simeq \varprojlim \tilde{K}_n(X_j) \oplus \varprojlim^1 \tilde{K}_{n+1}(X_j),$$

hence the groups Ext(X) are completely determined by K-homology theory.

5. Proof of the main theorem.

Our proof uses a spectral sequence which generalizes the Atiyah-Hirzebruch spectral sequence to compact metric spaces. The reader is referred to Kaminker - Schochet [7] for a detailed proof.

Theorem. Let X be a finite dimensional compact metric space, and h_* a generalized Steenrod homology theory. Then there is a spectral sequence converging to $h_*(X)$ with

$$E^2_{p,q} = {}^s\tilde{H}_p(X; h_q(S^0)).$$

Specialize to $E_* = h_*$ and $X \subseteq R^4$. Then the spectral sequence degenerates to yield the following diagram with exact row and columns:

$$\begin{array}{ccccccc}
 & 0 & & & & 0 & \\
 & \downarrow & & & & \uparrow & \\
 & {}^s\tilde{H}_1(X) & & & & {}^s\tilde{H}_2(X) & \\
 & \downarrow & & & & \uparrow & \\
 & Ext(X) & \overset{\Gamma_3}{\searrow} & & & E_0(X) & \\
 & \downarrow & & & & \uparrow & \\
0 \to & ker(d_3) & \longrightarrow {}^s\tilde{H}_3(X) & \xrightarrow{d_3} & {}^s\tilde{H}_0(X) \longrightarrow & cok(d_3) & \to 0 \\
 & \downarrow & & & & \uparrow & \\
 & 0 & & & & 0 &
\end{array} \tag{7}$$

It remains to show that d_3 is identically zero for all $X \subseteq R^4$ and that Γ_3 splits appropriately. First consider the case of a finite complex X. Then $Ext(X) \simeq K_1(X)$ by BDF [1] and $\tilde{K}_1(X) \simeq \tilde{H}_1(X) \oplus \tilde{H}_3(X)$ by an argument in homotopy theory. The isomorphism $Ext(X) \simeq \tilde{H}_1(X) \oplus \tilde{H}_3(X)$ thus obtained is natural for finite complexes in R^4.

In the general case, write $X = \varprojlim X_j$ where the X_j are finite complexes in R^4. Then there is a natural isomorphism

$$\check{E}_1(X) \simeq \check{\tilde{H}}_1(X) \oplus \check{\tilde{H}}_3(X) \tag{8}$$

obtained by taking limits. The $\varprojlim^1$ sequence (3) implies that the natural map ${}^s\tilde{H}_3(X) \to \tilde{H}_3(X)$ is an isomorphism for $X \subseteq R^4$, and the $\varprojlim^1$-sequence (6) implies that the natural map $\mathrm{Ext}(X) \to E_1(X)$ is onto. Now the diagram

$$\begin{array}{ccccccc} & & \mathrm{Ext}(X) & \longrightarrow & E_1(X) & \longrightarrow & 0 \\ & & \downarrow \Gamma_3 & & \downarrow & & \\ 0 & \longrightarrow & {}^sH_3(X) & \longrightarrow & \tilde{H}_3(X) & \longrightarrow & 0 \\ & & & & \downarrow & & \\ & & & & 0 & & \end{array} \qquad (9)$$

has exact rows and columns, and commutes. So the map Γ_3 is onto Contemplation of diagram (7) yields $d_3 \equiv 0$, completing the proof of the theorem.

6. Remarks.

Remark 1. If X does not separate R^4 then $\Gamma_1: \mathrm{Ext}(X) \longrightarrow {}^s\tilde{H}_1(X)$ is an isomorphism as previously observed (e.g., take X to be the torus $S^1 \times S^1$). What conditions on the operators A_1, A_2 would imply this? Similarly, what conditions on the operators would imply that

$$\sigma_e(A_1) \times \sigma_e(A_2) = \text{joint } \sigma_e(A_1, A_2)$$

(If this holds then Ext is much easier to calculate.)

Remark 2. The Chern character

$$\mathrm{ch}\otimes Q: \mathrm{Ext}(X)\otimes Q \longrightarrow H_{odd}(X;Q)$$

provides a means to systematically study torsion-free behavior, (where index information is classically to be found). This map is an isomorphism when

$$\hom(K^*(X), Q/Z) \otimes Q = 0 .$$

What (bizarre) phenomena in operator theory correspond to the non-vanishing of this group?

Remark 3. Let X be compact metric. Then Ext(X) is unnaturally isomorphic to the following sum:

$$\mathrm{Ext}(X) = \mathcal{L}E_1(X) \otimes \varprojlim (\text{torsion } (\tilde{K}_1(X_j)) \oplus \hom(\tilde{K}^1(X), Z) .$$

The first of these is divisible and uncountable and is the maximal divisible subgroup of Ext(X); the second is profinite; the third is a subgroup of $\Pi_j Z$, the Specker group. Which abelian groups may be realized as Ext(X)? We have partial results.

REFERENCES

1. L. G. Brown, R. G. Douglas, and P. A. Fillmore, Extensions of C*-algebras, operators with compact self-commutators, and K-homology, Bull. Amer. Math. Soc. 79 (1973), 973-978.

2. __________, Unitary equivalence modulo the compact operators and extensions of C -algebras, Proc. Conf. on Operator Theory, Lecture Notes in Math. vol. 345, Springer-Verlag, New York, 1973.

3. __________, Operator algebras and K-homology (to appear).

4. L. G. Brown, Operator algebras and algebraic K-theory (to appear).

5. L. G. Brown and C. Schochet, K_1 of the compact operators is zero, Proc. Amer. Math. Soc. (to appear).

6. S. Eilenberg and S. MacLane, Group extensions and homology, Ann. of Math. 43 (1942).

7. J. Kaminker and C. Schochet, Steenrod homology and operator algebras, Bull. Amer. Math. Soc. 81 (1975), 431-434.

8. __________, K-theory and Steenrod homology: applications to Brown-Douglas-Fillmore theory of operator algebras, Trans. Amer. Math. Soc. (to appear).

9. J. Milnor, On the Steenrod homology theory, Berkeley, 1961, (mimeographed).

10. N. Steenrod, Regular cycles on compact metric spaces, Ann. of Math. 41 (1940), 833-851.

ON ALGEBRAIC K-THEORY AND THE HOMOLOGY OF CONGRUENCE SUBGROUPS[1)]

Ronnie Lee and R. H. Szczarba, Yale University

The idea of defining the homology and cohomology of an algebraic system dates back to the work of Hurewicz [3]. He proved that the homology of a $K(\pi;1)$ space depends only on the group π. (A $K(\pi;1)$ space is one having trivial homotopy except for its fundamental group which is isomorphic to π.) Thus one can define the homology (or cohomology) groups of the group π to be the homology (or cohomology) of the any $K(\pi;1)$ space. Soon afterward, an algebraic definition of the homology and cohomology of a group was given which made possible the definition of the homology and cohomology of other algebraic systems.

In the late 50's, Atiyah, Hirzebruch and others developed an "extraordinary" cohomology theory for topological spaces, the so called K-theory. This theory proved to be extremely useful so it was natural to attempt to define K-groups for algebraic systems. The first step was suggested by the work of R. Swan [12]. He proved that, for any compact Hausdorf space X, $K^{o}(X)$ is isomorphic to the projective class group of the ring of complex values continuous functions on X. This suggested defining $K_{o}(\Lambda)$ to be the projective class group of Λ for any ring Λ. Subsequently, Bass [1] defined $K_1(\Lambda)$ and Milnor [8] defined $K_2(\Lambda)$ which fit naturally into this framework. Finally, Quillen [9] extended these definitions by defining $K_q(\Lambda)$ to be the homotopy groups of an appropriate space. (See section 1 below.) Up until this time, no purely algebraic definition has been given for $K_q(\Lambda)$ when $q > 2$.

Computations of the groups $K_q(\Lambda)$ has proved to be quite difficult. Quillen [11] has determined them when Λ is a finite field. If $\Lambda = Z$, the ring of integers, it is known that

During the preparation of this paper, both authors were partially supported by grant number NSF-MPS 74-08221.

$$K_o(Z) \simeq Z \quad ,$$
$$K_1(Z) \simeq Z_2 \quad ,$$
$$K_2(Z) \simeq Z_2 \quad .$$

(See Bass [1] and Milnor [8].) In addition, Quillen (unpublished)) has shown that $K_3(Z)$ contains $\pi_3 \simeq Z_{24}$, the stable 3-stem, and Karoubi [4] has shown that the order of $K_3(Z)$ must be at least forty eight. Our result is the following. (See [5] and [7].)

<u>THEOREM</u>: <u>The group $K_3(Z)$ is cyclic of order forty eight.</u>

In particular, π_3 is not a direct summand of $K_3(Z)$.

We note that our proof does not depend on the result of Karoubi mentioned above but does depend on the result of Quillen. We also note that Lichtenbaum has conjectured that the odd torsion subgroup of $K_q(Z)$ coincides with the odd torsion subgroup of the image of the J-homomorphism

$$J : \pi_q \; SO \longrightarrow \pi_q$$

when $q \equiv 3 \bmod 4$. Our result verifies the first case of this conjecture.

The purpose of this note is to outline a proof of the above theorm showing, in particular, how the homology of congruence subgroups enter into the argument. In the first section, we state some of the basic facts and give the general program. The second section deals with the odd torsion and the third with the even torsion. In the final section, we state our (far from complete) results on the homology of congruence subgroups.

1. The General Program.

Let Λ be a commutative ring with a unit and let $QP(\Lambda)$ be the category defined by Quillen [9]. The objects of $QP(\Lambda)$ are the finitely generated projective Λ-modules. A mapping from the object M to the object M' is a triple (φ, M_1, M_2) where $M_1 \subset M_2$ are projective submodules of M' and φ is an isomorphism of M onto

M_2/M_1 . Let $BQP(\Lambda)$ be the classifying space (or geometric realization) of the category $QP(\Lambda)$. Then $K_q(\Lambda)$ is defined to be the homotopy group $\pi_{q+1}BPQ(\Lambda)$.

We are concerned here with the ring Z of integers so we write Q for $QP(Z)$. In order to compute the low dimensional homotopy of BQ , we first compute the low dimensional homology of BQ . As we see below, this is sufficient to compute the odd torsion in $K_3(Z)$.

Let Q_n be the full subcategory of Q containing all free abelian groups of rank $\leq n$. According to Quillen [10], we have an exact sequence

(1.1) $\ldots \to H_q BQ_{n-1} \to H_q BQ_n \to H_{q-n}(GL_n(Z);St_n) \to H_{q-1}BQ_{n-1} \to H_{q-1}BQ_n \to \ldots$

where $GL_n(Z)$ is the group of $n \times n$ invertible integral matrices and St_n is the Steinberg module for the rational vector space of dimension n . Since

$$H_0(GL(n;Z);St_n) = 0$$

(See [6], Theorem 1.3), it follows that the mapping

$$H_q BQ_n \to H_q BQ$$

is an isomorphism for $q \leq n$ and an epimorphism for $q = n+1$. Now it is easy to show that BQ_1 has the same homotopy type as $S^1 \vee SRP^\infty$ so the determination of the low dimensional homology of BQ involves determining the groups $H_j(GL_n(Z);St_n)$ for low values of q and n .

For any odd prime p , we denote by F_p the field of p-elements and $\Gamma(n;p)$ the congruence subgroup of $SL_n(Z)$ at level p . Thus we have an exact sequence

(1.2) $$1 \to \Gamma(n;p) \to SL_n(Z) \to SL_n(F_p) \to 1 .$$

The groups $H_j(GL_n(Z);St_n)$ are easily obtained from the groups $H_j(SL_n(Z);St_n)$ and these groups are related to the groups

$$H_r(SL_n(F_p);H_s(\Gamma(n;p);St_n))$$

by the Hochschild-Serre spectral sequence of the exact sequence above. This leads to the problem of computing the homology of $\Gamma(n;p)$ as a module over $SL_n(F_p)$.

Let $X = SL_n(R)/SO(n)$. Then $\Gamma = \Gamma(n;p)$ acts freely on X and the homology of Γ is isomorphic to the homology of the locally symmetric manifold $M = \Gamma \backslash X$. Now M is not compact but does have finite volume (as a locally symmetric manifold). We can then compactify M obtaining a manifold $\bar{M}$ with boundary whose interior is M . In fact, $\bar{M}$ is obtained by enlarging X to a manifold $\bar{X}$ with "corners" on which Γ acts freely so that $\bar{M} = \Gamma \backslash \bar{X}$. (See Borel-Serre [2].) In the case $n = p = 3$, we are able to describe $\partial\bar{M}$ in detail and use this description to determine the homology and cohomology of $\Gamma(3;3)$ as a module over $SL_3(F_3)$ (See [6].) This in turn allows us to compute the low dimensional homology of BQ .

2. The Odd Torsion in $K_3(Z)$.

We now sketch the proof of the following (See [7].):

THEOREM 2.1: <u>The odd torsion subgroup of $K_3(Z)$ is cyclic of order three.</u>

We begin by reducing this computation to one in homology.

LEMMA 2.1: <u>The odd torsion subgroup of $\pi_4 BQ$ is isomorphic to the odd torsion subgroup of $H_4 BQ$.</u>

Proof:

Let $\pi : \tilde{B} \to BQ$ be the universal covering and $\alpha : S^1 \to BQ$ a continuous mapping representing a generator of $\pi_1 BQ \simeq Z$. Using the fact that BQ is an H-space (in fact, an infinite loop space), we can extend the mapping

$$\pi \vee \alpha : \tilde{B} \vee S^1 \to BQ$$

to a mapping $\varphi : \tilde{B} \times S^1 \to BQ$ which clearly induces isomorphisms on homotopy. Thus BQ has the same homotopy type as $\tilde{B} \times S^1$. Lemma 2.2 now follows from the mod C Hurewicz Theorem, the Kunneth Theorem, and the fact (See Quillen [11]) that

$$\pi_1\widetilde{B} = 0 ,$$
$$\pi_2\widetilde{B} \simeq \pi_2 BQ \simeq Z_2 ,$$
$$\pi_3\widetilde{B} \simeq \pi_3 BQ \simeq Z_2 .$$

As we saw in section 1, the homomorphism

$$H_4 BQ_3 \to H_4 BQ$$

is surjective. In addition, we know that $\pi_4 BQ$ has order at least 24 so $\pi_4 BQ$ and thus $H_4 BQ$ must contain a cyclic group of order three. Thus, Theorem 2.1 will be proved if we can show that the odd torsion subgroup of $H_4 BQ_3$ is cyclic of order three. To do this, we consider the exact sequence (1.1) for small values of n and q :

$$H_4 BQ_1 \to H_4 BQ_2 \to H_2(GL_2(Z);St_2) ,$$
$$H_4 BQ_2 \to H_4 BQ_3 \to H_1(GL_3(Z);St_3) .$$

As mentioned above, BQ_1 has the same homotopy type as $S^1 \vee SRP^\infty$ so $H_4 BQ_1 \simeq Z_2$. Furthermore, we can prove

LEMMA 2.3: <u>The group $H_1(GL_3(Z);St_3)$ is trivial and $H_2(GL_2(Z);St_2)$ is cyclic of order 12 .</u>

Proof:

The isomorphism $H_2(GL_2(Z);St_2) \simeq Z_4 \oplus Z_3$ follows from the fact that St_2 can be explicitly described. (See [7], section 3.) We now sketch a proof of the fact that $H_1(GL_3(Z);St_3)$ is trivial. We will, for convenience, denote $GL_3(Z)$ by GL , $GL_3(F_3)$ by $\overline{GL}$, $\Gamma(3;3)$ by Γ , and St_3 by St . In addition, we let $\overline{St}$ be the Steinberg module for the three dimensional vector space over the field of three elements.

Associated with the extension

$$1 \to \Gamma \to GL \to \overline{GL} \to 1$$

is the Hochschild-Serre exact sequence

$$H_2(GL;St) \to H_2(\overline{GL};H_o(\Gamma;St)) \to H_o(\overline{GL};H_1(\Gamma;St))$$
$$\to H_1(GL;St) \to H_1(\overline{GL};H_o(\Gamma;St)) \to 0 .$$

The Borel-Serre duality (See [2].) implies that $H_1(\Gamma;St)$ is isomorphic to $H^2(\Gamma)$ and Theorem 1.2 of [6] states that

$$H_o(\Gamma;St) \simeq \overline{St} .$$

Thus the sequence above becomes

$$H_2(GL;St) \to H_2(\overline{GL};\overline{St}) \to H_o(\overline{GL};H^2(\Gamma))$$
$$\to H_1(GL;St) \to H_1(\overline{GL};\overline{St}) \to 0 .$$

Further computation shows that $H_1(\overline{GL};\overline{St}) = 0$ (See [7], Section 4), $H_o(\overline{GL};H^2(\Gamma)) \simeq Z_2$ (See [6], Lemma 12.5), and the homomorphism

$$H_2(GL;St) \to H_2(\overline{GL};\overline{St})$$

is <u>not</u> surjective (See Lemma 4.1 of [7]). It follows that $H_1(GL;St)$ is trivial.

3. The 2-torsion Subgroup of $K_3(Z)$.

The determination of the 2-torsion subgroup of $K_3(Z)$ is far more difficult than the odd torsion. In this section, we give a rough sketch of this determination. (See [7] for the detailed argument.)

Let $\Omega^{\infty-1}S^\infty = \underset{\to}{\lim}\,\Omega^{n-1}S^n$. Then $\pi_{q-1}(\Omega^{\infty-1}S^\infty) \simeq \pi_q$ the stable q-stem of the homotopy groups of spheres. One can define a mapping

$$f : \Omega^{\infty-1}S^\infty \to BQ$$

(See, for example, section 1 of [7]) which induces homomorphisms

$$f_* : \pi_q\Omega^{\infty-1}S^\infty \to \pi_q BQ \simeq K_{q-1}(Z) .$$

In fact, f_* is an isomorphism for $q < 3$ (See Bass [1] and Milnor [8]) and a monomorphism for $q = 3$ (Quillen, unpublished).

Let F be the homotopy fibre of f . Then F is 2-connected and we have a short exact sequence

$$0 \to \pi_4\Omega^{\infty-1}S^\infty \to \pi_4 BQ \to \pi_3 F \to 0 . \tag{3.1}$$

Since $\pi_4\Omega^{\infty-1}S^\infty \simeq \pi_4 \simeq Z_{24}$, our main theorem is a consequence of the following:

THEOREM 3.1: <u>The group $\pi_3 F$ is cyclic of order two and the sequence (3.1) does not split.</u>

The idea of the proof of this result to compare the space BQ with the space $BQP(F_3) = B\bar{Q}$ which is well understood by results of Quillen [11]. In particular, we have a commutative diagram

$$\begin{array}{ccc} \Omega^{\infty-1}S^{\infty} & \xrightarrow{f} & BQ \\ & \searrow^{h} & \downarrow g \\ & & B\bar{Q} \end{array}$$

where $g : BQ \to B\bar{Q}$ is induced by the homomorphism $Z \to F_3$ and $h = gf$. Consideration of the Seere exact sequence of the fibration $F \to \Omega^{\infty-1}S^{\infty} \to BQ$ leads to an exact sequence

$$H_4\Omega^{\infty-1}S^{\infty} \to H_4BQ \to H_3F \to 0 .$$

Combining this sequence with the diagram above, we obtain

$$(3.2) \qquad \begin{array}{ccccccc} H_4\Omega^{\infty-1}S^{\infty} & \xrightarrow{f_*} & H_4BQ & \longrightarrow & H_3F & \longrightarrow & 0 \\ & \searrow^{h_*} & \downarrow g_* & & & & \\ & & H_4B\bar{Q} . & & & & \end{array}$$

Using Theorem 2.1 and known facts about the space $B\bar{Q}$ and the mapping h, we are able to prove that, modulo odd torsion,

$$H_4BQ \simeq Z_4 \oplus Z_2 ,$$

$$H_4B\bar{Q} \simeq Z_4 \oplus Z_4 ,$$

$$g_* : H_4BQ \to H_4B\bar{Q} \text{ is injective,}$$

and the image of $h_* : H_4\Omega^{\infty-1}S^{\infty} \to H_4B\bar{Q}$ is a subgroup of index four. (See [7]). It now follows that

$$\pi_4F \simeq H_4F \simeq Z_2 .$$

To see that π_3F is not a direct summand of π_4BQ, we consider the diagram

$$\begin{array}{ccccccccc} 0 & \longrightarrow & \pi_4\Omega^{\infty-1}S^{\infty} & \xrightarrow{f_*} & \pi_4BQ & \longrightarrow & \pi_3F & \longrightarrow & 0 \\ & & & \searrow^{h_*} & \downarrow g_* & & & & \\ & & & & \pi_4B\bar{Q} . & & & & \end{array}$$

It is known that the image of the homomorphism

$$h_* : \pi_4 \Omega^{\infty-1} S^\infty \to \pi_4 B\bar{Q}$$

is a subgroup of index two (See Quillen [11].) and we prove that the homomorphism

$$g_* : \pi_4 BQ \to \pi_4 B\bar{Q}$$

is surjective. (See [7].) Since $\pi_4 B\bar{Q}$ is cyclic of order eight, it follows that $\pi_3 F$ cannot be a direct summand of $\pi_4 BQ$.

4. The Homology of $\Gamma(n;p)$.

Let $\Gamma(n;p)$ be the congruence subgroup of $SL_n(Z)$ of level p . It follows from the exact sequence (1.2) above that $H_*(\Gamma(n;p))$ is a module over $SL_n(F_p)$. We now state our results on the determination of this module. (See [5] and [6].) These results are complete only when $n = p = 3$.

Let $M(n;p)$ be the abelian group of $n \times n$ matrices with entries in F_p and $M_o(n;p)$ the subgroup of matrices with zero trace. The group $SL_n(F_p)$ acts on the right on $M(n;p)$ and $M_o(n;p)$ by

$$Ag = g^{-1}Ag$$

for $A \in M(n;p)$ and $g \in SL_n(F_p)$. Define $\varphi : \Gamma(n;p) \to M(n;p)$ by

$$\varphi(A) = \frac{1}{p}(A-I) \bmod p \ .$$

THEOREM 4.1: <u>The mapping $\varphi : \Gamma(n;p) \to M(n;p)$ is a homomorphism. Furthermore, for any prime p and integer $n \geq 3$, φ defines an isomorphism of $H_1(\Gamma(n;p))$ onto $M_o(n;p)$ as modules over $SL_n(F_p)$.</u>

The proof of this result uses the solution to the congruence subgroup problem.

For any integer $n \geq 2$ and prime $p > 2$, we let $St_n(p)$ denote the Sternberg module over $SL_n(F_p)$ for the vector space of dimension n over F_p .

THEOREM 4.2: <u>For any integer $n \geq 3$, $H_o(\Gamma(n;3);St_n)$ is isomorphic to $St_n(3)$ as modules over $SL_n(F_p)$.</u>

The Borel-Serre duality theorem states that for any $q \geq 0$, $n \geq 2$, and odd prime p,

$$H^q(\Gamma(n;p)) \simeq H_{N-q}(\Gamma(n;p);St_n)$$

as modules over $SL_n(F_p)$ where $N = \frac{1}{2}(n^2 - n)$. (See [2].) As a consequence of this result and the universal coefficient theorem, we have the

COROLLARY: <u>For any integer $n \geq 3$, $H^N(\Gamma(n;3))$ is isomorphic to $St_n(3)$ and $H_N(\Gamma(n;3))$ is isomorphic to $Hom(St_n(3);Z)$ as modules over $SL_n(F_3)$.</u>

The proof of Theorem 4.2 involves an explicit free resolution of St_n over $SL_n(Z)$.

To state our final result, we let P_1 and P_2 be two non conjugate maximal parabolic subgroups of $SL_3(F_3)$. Let

$$\rho_1, \rho_2 : SL_3(F_3) \to GL_{13}(Z)$$

be the representations obtained by inducing up from the unique non-trivial one dimensional representations

$$P_1, P_2 \to GL_1(Z) .$$

Finally, let A and B be the $SL_3(F_3)$ modules defined by ρ_1 and ρ_2 respectively.

THEOREM 4.3: <u>There is a short exact sequence</u>

$$0 \to A \oplus B \to H_2(\Gamma(3;3)) \to Z_2 \to 0$$

<u>of $SL_3(F_3)$ modules.</u>

REFERENCES

1. H. Bass, K-theory and stable algebra, Publ. I.H.E.S. No. 22, (1964), 5-60.

2. A. Borel and J-P. Serre, Corners and arithmetic groups, Comm. Math. Helv., 48 (1973), 436-491.

3. W. Hurewicz, Beiträge zur Topologie der Deformationen IV; Asphärische Raüme, Proceedings, Akademie van Wetenschappen, Amsterdam, 39 (1936), 215-224.

4. M. Karoubi, Périodicité de la K-Théorie Hermitienne, Springer Lecture Notes in Mathematics Series, vol. 343, 301-411.

5. R. Lee and R. H. Szczarba, On the homology of congruence subgroups and $K_3(Z)$, Proc. Nat. Acad. Sci. U.S.A., 72 (1975), 651-653.

6. R. Lee and R. H. Szczarba, On the homology and cohomology of congruence subgroups (to appear).

7. R. Lee and R. H. Szczarba, On $K_3(Z)$ (submitted for publication).

8. J. Milnor, Introduction to algebraic K-theory, Annals of Math. Studies 72, 1971.

9. D. Quillen, Higher algebraic K-theory I, Springer Lecture Notes in Mathematics Series, vol. 341, 84-147.

10. D. Quillen, Finite generation of the groups K_i of rings of algebraic integers, Springer Lecture notes in Mathematics Series, vol. 341, 179-198.

11. D. Quillen, On the cohomology and K-theory of the general linear group over a finite field, Ann. of Math., 96 (1972), 552-586.

12. R. W. Swan, Vector bundles and projective modules, Trans. A.M.S., 105 (1962), 265-277.

CONFIGURATION SPACES

Dusa MacDuff, Massachusetts Institute of Technology

§1. Introduction.

This a a talk about configuration spaces and as such is not directly relevant to the main theme of the conference. However it should illustrate the kind of topological reasoning which lies behind some of the results mentioned in Segal's talk on algebraic K-theory. In particular I will sketch a proof of the Barratt-Quillen-Priddy theorem that $B\Sigma_\infty$ is homology isomorphic to $(\Omega^\infty S^\infty)_0$, and also give a simpler formulation of the Atiyah-Singer proof of the Bott periodicity theorem in [1], incidentally removing from it all the analysis.

The configuration space $C(M)$ of a smooth manifold with boundary M is the set of finite subsets of M (think of a finite subset of M as a configuration of particles on M) topologised so that the particles cannot collide. Thus $C(M)$ is the disjoint union $\coprod_{k\geq 0} C_k(M)$, where $C_k(M)$ is the configurations with k particles, i.e. the quotient of the ordered configuration space $OC_k(M) = \{(m_1,\ldots,m_k): m_i \in M,$ $m_i \neq m_j$ if $i \neq j\}$ by the action of the symmetric group Σ_k. The configuration space $C_k(\mathbb{R}^n)$ is an approximation to the classifying space $B\Sigma_k$. For $OC_k(\mathbb{R}^n)$ is just $\mathbb{R}^{nk}$ with certain hyperplanes of codimensionnn removed, so its homotopy groups vanish up to dimension $n-2$. Thus $\lim_{n\to\infty} OC_k(\mathbb{R}^n)$, where the limit is taken with respect to the usual inclusions $\mathbb{R}^n \longrightarrow \mathbb{R}^{n+1}$, is a contractible space on which Σ_k acts freely. The quotient $\lim_{n\to\infty} C_k(\mathbb{R}^n)$ is therefore homotopic to $B\Sigma_k$.

Now, any finite subset $s \subseteq \mathbb{R}^n$ gives rise to a map $\varphi(s): \mathbb{R}^n \longrightarrow \mathbb{R}^n \cup \{\infty\}$. For let $\{\delta_x\}_{x\in s}$ be a collection of disjoint open discs with centres at the points x of s and with radius $\epsilon(x)$ (choose $\epsilon(s) > 0$ so that it is as large as possible but ≤ 1 say). Then define $\varphi(s)$ by:

$$\varphi(x)y = \infty \qquad \text{if } y \text{ is not in any } \delta_x,$$

$$\varphi(s)y = (\epsilon(x)-\|y-x\|)^{-1}(y-x) \text{ if } y \text{ is in } \delta_x.$$

This $\varphi(s)$ actually has compact support (in the sense that $\varphi^{-1}(\mathbb{R}^n)$ has compact closure), so φ maps $C_k(\mathbb{R}^n)$ to $\mathrm{Map}_k(\mathbb{R}^n, \mathbb{R}^n \cup \{\infty\})$, the space of maps $\mathbb{R}^n \longrightarrow \mathbb{R}^n \cup \{\infty\}$ with compact support and degree k. Notice that the spaces $\mathrm{Map}_k(\mathbb{R}^n, \mathbb{R}^n \cup \{\infty\})$ are homotopic for different k, and, in fact, a e all homotopic to $(\Omega^n S^n)_0$, the space of base-point preserving maps with degree 0 from the n-sphere S^n to itself (identifying S^n with $\mathbb{R}^n \cup \{\infty\}$). Also, we can form $\lim\limits_{k\to\infty} C_k(\mathbb{R}^n)$ by mapping $C_k(\mathbb{R}^n)$ to $C_{k+1}(\mathbb{R}^n)$ by adding a particle from infinity in a standard way. Altogether we get a map $\varphi: \lim\limits_{k\to\infty} C_k(\mathbb{R}^n) \longrightarrow (\Omega^n S^n)_0$, and this, by a theorem of Segal [6], induces an isomorphism on integral homology. Taking the limit now over n gives the Barratt-Quillen-Priddy theorem that $\lim\limits_{n,k\to\infty} C_k(\mathbb{R}^n) \simeq \lim\limits_{k\to\infty} B\Sigma_k \longrightarrow (\Omega^\infty S^\infty)_0$ is a homology isomorphism, or equivalently that $(B\Sigma_\infty)^{ab} \longrightarrow (\Omega^\infty S^\infty)_0$ is a homotopy equivalence, where ab denotes Quillen's construction abelianising the fundamental group.

Segal's theorem is a special case of a more general theorem valid for any connected manifold M. Let $\Gamma_k(M)$ be the space of possibly infinite, compactly supported vector fields on M, that is, the space of sections of the fibre bundle over M whose fibre at $x \in M$ is the sphere $T_x \cup \{\infty\}$ obtained by compactifying the tangent space T_x at x. Then, as before, there is a map $\varphi: C_k(M) \longrightarrow \Gamma_k(M)$. Because it is always possible to move the particles away from the boundary ∂M of M, φ may be defined so that it takes values in $\Gamma_k(M, \partial M)$, the sections which equal ∞ on ∂M.

<u>Theorem 1</u>. $\varphi: C_k(M) \longrightarrow \Gamma_k(M, \partial M)$ induces an isomorphism on integral homology groups up to a dimension tending to ∞ with k.

Taking the limit as $k \longrightarrow \infty$ gives the previous result when $M = \mathbb{R}^n$. Notice that because homology commutes with direct limits it is enough to prove this for compact M. Also, the result for $\mathbb{R}^n$ follows immediately from that for D^n, where D^n is the unit disc in $\mathbb{R}^n$, since $C(\mathbb{R}^n)$ is homotopic to $C(D^n)$. Therefore from now on we will consider only compact manifolds.

§2. Gromov's method. Theorem 1 is proved by a very general method which was first used systematically by Gromov [3]. Let $\mathcal{M}$ be the category of smooth, compact manifolds of a fixed dimension n with embeddings as morphisms, and suppose that F is a contravariant functor from $\mathcal{M}$ to (spaces). Associated to F there is a fibre bundlo $E_F(M)$ on each $M \in Obj(\mathcal{M})$ which has fibre $F(D_x)$ at the point x of M. (Here M is supposed to have a Riemannian metric, and D_x is the unit disci n the tangent space to M at x, so that $E_F(M)$ is associated to the tangent bundle on M.) The exponential map gives an embedding $exp_x : D_x \longrightarrow M$ for each x, and one defines $\varphi : F(M) \longrightarrow \Gamma_F(M)$ (where $\Gamma_F(M)$ is the space of continuous sections of $E_F(M)$) by $\varphi(f)(x) = F(exp_x)f$, for f in F(M). Then we have

Theorem 2. Suppose that

(i) for all embeddings $j : D^n \longrightarrow D^n$, $F(j) : F(D^n) \longrightarrow F(D^n)$ is a homotopy equivalence;

(ii) for any square

$$\begin{array}{ccc} M_1 \cup M_2 & \longleftarrow & M_1 \\ \uparrow & & \uparrow \\ M_2 & \longleftarrow & M_1 \cap M_2 \end{array}$$

in $\mathcal{M}$ the corresponding square

$$\begin{array}{ccc} F(M_1 \cup M_2) & \longrightarrow & F(M_1) \\ \downarrow & & \downarrow \\ F(M_2) & \longrightarrow & F(M_1 \cap M_2) \end{array}$$

is homotopy Cartesian.

Then $\varphi : F(M) \longrightarrow \Gamma_F(M)$ is a homotopy equivalence for all $M \in Obj(\mathcal{M})$. (A square

$$\begin{array}{ccc} W & \longrightarrow & X \\ \downarrow & & \downarrow f \\ Z & \xrightarrow{g} & Y \end{array}$$

is called Cartesian if W is the fibre product $\{(x,y) ; f(x) = g(z)\}$ of X and Z over Y. Similarly it is called homotopy Cartesian if W is homotopic to the homotopy-theoretic fibre product of X and Z over Y, that is if $W = \{(x,z,\gamma) : \gamma$ is a path in Y with $\gamma(0) = f(x)$, $\gamma(1) = g(z)\}$.)

The proof of this theorem is trivial. For it follows immediately from (i) that it holds when $M = D^n$. The general case follows by induction over the number of discs in a finite covering of M, because condition (ii), which holds also for the functor Γ_M, implies that if φ is an equivalence for $M_1 \cap M_2$, M_1 and M_2 it is one for $M_1 \cup M_2$ too.

There is another version of this theorem where the square in (ii) satisfy the condition that no component of $\partial(M_1 \cap M_2)$ lies entirely in the interior of M_1. In this case one argues by induction over the number of handles in a handle decomposition of M, where the handles have index < dim(M), and proves that φ is an equivalence for all open M, i.e. for M for which $M - \partial M$ has no compact component.

The interest of this theorem lies in the fact that there are many functors F satisfying these conditions. An example of the kind considered by Gromov in his thesis [3] is the functor which assigns to each manifold the space of all its symplectic structures, and the theorem yields an existence statement: and open manifold M of dimension 2k has a symplectic structure iff the obvious algebraic condition is satisfied, namely, if there is some (not necessarily closed or smooth) 2-form ρ on M such that ρ^k has no zeroes.

To apply Gromov's theorem one needs a criterion for a square to be homotopy Cartesian. A well-known result is

Lemma. A Cartesian square $\begin{array}{ccc} W & \longrightarrow & X \\ & & \downarrow f \\ Z & \xrightarrow{g} & Y \end{array}$ is homotopy Cartesian if f is a fibration.

Proof. We must show that the inclusion of $W = \{(x,z): f(x) = g(z)\}$ in $W' = \{(x,z,\gamma): \gamma$ is a path in Y with $\gamma(0) = f(x)$, $\gamma(1) = g(z)$ is a homotopy equivalence. However, in order to retract W' to W it is enough to lift the homotopy $W' \times I \longrightarrow Y: ((x,z,\gamma),t) \longrightarrow \gamma(t)$ to X with initial lifting $((x,z,\gamma),0) \longrightarrow x$, and this may be done since f is a fibration.

In fact it suffices here that f be a quasifibration, that is a map such that for all $y \in Y$ the inclusion of the actual fibre $f^{-1}(y)$ at y into the homotopy fibre $F(y,f) = \{(x,\gamma):\gamma$ is a path in Y with $\gamma(0) = y$, $\gamma(1) = f(x)\}$ is an equivalence. For the purposes of homotopy theory these maps are just as good as fibrations, and for instance have a long exact homotopy sequence. They often arise in the following form. The base space Y is filtered by an increasing sequence of closed subspaces $Y_1 \subseteq Y_2 \subseteq Y_3 \cdots$ such that, over each difference $Y_k - Y_{k-1}$, f is the product fibration $(Y_k - Y_{k-1}) \times F \longrightarrow (Y_k - Y_{k-1})$. (Notice that this means that all the fibres $f^{-1}(y)$ are homeomorphic to F.) Also for each k there is an open neighborhood $\mathcal{U}_k$ of Y_k in Y_{k+1} and a deformation retraction r_t of $\mathcal{U}_k$ to Y_k which may be lifted to a deformation retraction $\tilde{r}_t$ of $f^{-1}(\mathcal{U}_k)$ to $f^{-1}(Y_k)$. Since all the fibres may be identified with F the maps $\tilde{r}_1 : f^{-1}(y) \longrightarrow f^{-1}(r_1(y))$ give rise to a collection of maps $F \longrightarrow F$, called the attaching maps of f. If they are all homotopy equivalences f is a quasifibration (see [2]). If they are all homology equivalences f can be called a "homology fibration" (see [4,5]).

§3. The application to configuration spaces.

In order to apply all this to configuration spaces we must introduce a new functor of the correct variance, for $M \longrightarrow C(M)$ is obviously covariant. Thus we consider $\tilde{C}(M)$, the configuration space of particles on M which are annihilated and created on the boundary of M. More formally, $\tilde{C}(M)$ is the quotient of $C(M)$ by the relation $s \sim s'$ iff $s \cap (M - \partial M) = s' \cap (M - \partial M)$. Clearly any embedding $N \longrightarrow M$ gives rise to a restriction map $\tilde{C}(M) \longrightarrow \tilde{C}(N) : s \longrightarrow s \cap N$, so that $\tilde{C}$ is a contravariant functor.

It is not difficult to see that $\tilde{C}(D^n)$ is homotopic to S^n. For, by expanding radially from the centre of D^n, it is possible to push all but at most one particle in each configuration out to the boundary where it vanishes. This retracts $\tilde{C}(D^n)$ to its subspace consisting of configurations of at most one particle, and this is just $D^n/\partial D^n \simeq S^n$. This argument also shows that $\tilde{C}$ satisfies condition (i) of Theorem 2. Notice, too, that the fibre $\tilde{C}(D_x)$ of the bundle $E_{\tilde{C}}(M)$ is $D_x/\partial D_x \cong T_x \cup \{\infty\}$, so that the elements of $\Gamma_{\tilde{C}}(M)$ may be considered to be possibly infinite vector fields on M.

Now consider the restriction map $r : \tilde{C}(M) \longrightarrow \tilde{C}(N)$ induced by an inclusion $N \longrightarrow M$. Filter $\tilde{C}(N)$ by the sets $\tilde{C}_k$ consisting of $\leq k$ particles. Then $r^{-1}(\tilde{C}_k - \tilde{C}_{k-1})$ is just the product $(\tilde{C}_k - \tilde{C}_{k-1}) \times F$, where F is the space of configurations in the closure $\overline{M-N}$ of M-N which are annihilated on $\partial(\overline{M-N}) \cap \partial M$ but not the rest of $\partial(\overline{M-N})$. We may choose $\mathcal{U}_k \subseteq \tilde{C}_{k+1}$ to consist of configurations with at least one particle near ∂N, so that $\mathcal{U}_k$ retracts to $\tilde{C}_k$ by pushing this particle out to ∂N. The attaching map on the fibre is then just the map $F \longrightarrow F$ which adds a particle to each configuration at some point m on $\partial(\overline{M-N}) \cap N$. Clearly this map will not be an equivalence if $\partial(\overline{M-N})$ lies entirely in the interior of M since F is then $C(\overline{M-N})$, a configuration space with no annihilations. However it is an equivalence if each component of ∂N meets ∂M, for then the added particle may be moved along near ∂N until it reaches ∂M where it disappears. (Notice that we can assume that all particles except the extra one have been cleared away from a neighborhood of $\partial N \cap (\overline{M-N})$ so that no collisions will occur.)

Thus the functor $\tilde{C}$ satisfies the (modified) condition (ii) in Theorem 2. It follows that $\varphi:\tilde{C}(M) \longrightarrow \Gamma_{\tilde{C}}(M)$ is an equivalence for all connected manifolds M with non-empty boundary. In order to obtain a theorem about C(M) one adds an annulus $\partial M \times I$ to M along ∂M and considers the commutative diagram

$$\begin{array}{ccc} \tilde{C}(M \cup_{\partial M} \partial M \times I) & \xrightarrow[\simeq]{\varphi} & \Gamma_{\tilde{C}}(M \cup_{\partial M} \partial M \times I) \\ \tilde{C}(\partial M \times I) & \xrightarrow[\simeq]{\varphi} & \Gamma_{\tilde{C}}(\partial M \times I) \end{array}$$

The actual fibres of the restriction maps r and R are $C(M) = \coprod_{k \geq 0} C_k(M)$ and $\Gamma_{\tilde{C}}(M, \partial M)$. They are not equivalent because, although

R is a fibration, as we saw above r is not a quasifibration. However, by "stablising the fibre with respect to the attaching maps" one alters r to a map $r':X \longrightarrow \tilde{C}(\partial M \times I)$ with fibre $Z \times \lim_{k\to\infty} C_k(M)$, where this limit is formed with respect to the attaching maps, that is the maps which add a particle to the configurations in $C_k(M)$ at some point m on ∂M. The attaching maps of r' are essentially the same as those of r, but now they are homology isomorphisms. This implies that r' is a homology fibration in the sense mentioned above. Thus φ induces a homology isomorphism between the fibres of r' and R. Theorem 1 now follows. The details of this argument may be found in [4].

§4. Bott periodicity.

I shall conclude by describing a variant of the Atiyah-Singer proof [1] of the Bott periodicity theorem to show how closely it is related to the preceding argument. Let U_∞ be the stable unitary group $\lim_{n\to\infty} U_n$. We shall construct a quasifibration $f:H \longrightarrow U_\infty$ with H contractible and fibre $Z \times BU_\infty$. This is enough to prove the complex Bott periodicity theorem. The real case can be treated similarly.

Let H_n be the $n \times n$ Hermitian matrices with all eigenvalues in $[0,1]$ (so H_n is linearly contractible) and define $f_n:H_n \longrightarrow U_n$ by $f_n(h) = \exp(2\ ih)$. The fibre $f_n^{-1}(u)$ of f_n at $u \in U_n$ may be identified with the grassmannian of all subspaces of $\ker(u-1)$ by the map $h \longrightarrow \ker(h-1) \subseteq \ker(u-1)$. The f_n are compatible with the usual inclusions $H_n \longrightarrow H_{n+1}$ and $U_n \longrightarrow U_{n+1}$ (given by $h \longrightarrow h\oplus 0$ and $u \longrightarrow u\oplus 1$) so that one has $f_\infty:H_\infty \longrightarrow U_\infty$ with all fibres homeomorphic to $\coprod_{n\geq 0} Gr_n(C^\infty) \simeq \coprod_{n\geq 0} BU_n$. Filtering the base U_∞ by the U_n one sees that the attaching maps come from inclusions $BU_m \longrightarrow BU_{n+r}$. All one needs to do now is to stabilise so that the fibres become $Z \times BU_\infty$ and the attaching maps homotopy equivalences. To do this, take the standard $V = C^\infty$ with basis $\{e_i\}_{-s<i<\infty}$. Let V_k be the subspace spanned by e_i for $i < k$. Then $Z \times BU_\infty$ can be thought of as the space of all subspaces W of V such that $V_p \subset W \subset V_q$ for some $-\infty < p \leq q < \infty$. Define $k_0 : V \longrightarrow V$ by

$$k_0(e_i) = e_i \qquad (i<0)$$
$$= 0 \qquad (i\geq 0),$$

and let H be the Hermitian operators $V \longrightarrow V$ which have eigenvalues in $[0,1]$ and have the form $k_0 + k$, where k is represented with respect to the basis $\{e_i\}$ by a matrix with finitely many non-zero terms. Similarly let U be the unitary operators $V \longrightarrow V$ of the form $I + v$, where the matrix for v has finitely many non-zero terms. Define $f : H \longrightarrow U$ by $h \longrightarrow \exp(2 ih)$. Then for each $u \in U$ $f^{-1}(u) \simeq Z \times BU_\infty$, where the identification is as above but with V replaced by its subspace $\ker(u-1)$. The attaching maps arise from inclusions $\ker(u-1) \longrightarrow \ker(u'-1)$. Thus they are equivalences, and f is a quasifibration.

REFERENCES

1. M.F. Atiyah and I.M. Singer, Index theory for skew-adjoint Fredholm operators, I.H.E.S. 37(1969), 5-26.

2. A. Dold and R. Thom, Quasifaserungen und unendliche symmetrische produkte, Ann. of Math. 67(1958), 239-281.

3. M.I. Gromov, Izv. Akad. Nauk SSSR 33(1969), 707-734.

4. D.McDuff, Configuration spaces of positive and negative particles, Topology 14(1975), 91-107.

5. D.McDuff and G.B. Segal, Homology fibrations and the "group completion" theorem, to appear.

6. G.B. Segal, Configuration spaces and iterated loop spaces, Invent. Math. 21(1973), 213-221.

EXTENSIONS OF C^*-ALGEBRAS AND THE REDUCING ESSENTIAL MATRICIAL SPECTRA OF AN OPERATOR

Carl Pearcy, University of Michigan and Norberto Salinas, University of Kansas

This is an expository paper whose purpose is to survey the results of the authors papers [10] and [11] and to show the relevance of these results to the problem of generalizing the recent Brown-Douglas-Fillmore theorem [4] that classifies essentially normal operators.

In this paper $\mathcal{H}$ will always be a separable, infinite dimensional, complex Hilbert space, and $\mathcal{L}(\mathcal{H})$ will denote the algebra of all bounded linear operators on $\mathcal{H}$. The ideals in $\mathcal{L}(\mathcal{H})$ of compact operators and trace-class operators will be denoted, respectively, by $\mathbb{K}(\mathcal{H})$ and $J(\mathcal{H})$. The spectrum of an operator T in $\mathcal{L}(\mathcal{H})$ will be denoted by $\sigma(T)$, the canonical quotient map of $\mathcal{L}(\mathcal{H})$ onto the Calkin algebra $\mathcal{L}(\mathcal{H})/\mathbb{K}(\mathcal{H})$ by π, and the [left, right] essential spectrum of T (i.e., the [left, right] spectrum of $\pi(T)$) by $[\sigma_{\ell e}(T), \sigma_{re}(T)]$ $\sigma_e(T)$. Recall that an operator T in $\mathcal{L}(\mathcal{H})$ is hyponormal [essentially hyponormal, essentially normal] if $TT^* \leq T^*T$ $[\pi(TT^*) \leq \pi(T^*T), \pi(TT^*) = \pi(T^*T)]$.

DEFINITION 1 (cf. [6]). Operator T_1 and T_2 acting on separable Hilbert spaces $\mathcal{H}_1$ and $\mathcal{H}_2$, respectively, are said to be <u>compalent</u> (notation: $T_1 \sim T_2$) if there exist a unitary operator $U : \mathcal{H}_1 \to \mathcal{H}_2$ and a compact operator K in $\mathbb{K}(\mathcal{H}_2)$ such that $UT_1U^* + K = T_2$. If, moreover, $\|K\| < \epsilon$, then T_1 and T_2 are <u>ϵ-compalent</u> (notation: $T_1 \sim T_2(\epsilon)$). If, on the other hand, $K \in J(\mathcal{H})$, then T_1 and T_2 are <u>J-compalent</u> (notation: $T_1 \sim T_2(J)$), and in case $\|K\|_J < \epsilon$, then T_1 and T_2 are <u>J_ϵ-compalent</u> (notation: $T_1 \sim T_2(J_\epsilon)$.

It is clear that compalence and J-compalence are equivalence relations, and in the past few years several interesting and important results have been obtained concerning the question of when two operators are compalent. Recall that a (normal) operator N in $\mathcal{L}(\mathcal{H})$ is said to be <u>diagonable</u> if there exists an orthonormal basis $\{e_n\}_{n=1}^{\infty}$ for $\mathcal{H}$ such that the matrix of N relative to this basis is diagonal.

THEOREM 1 (Berg [2], Halmos [5]). If N is a normal operator in $\mathcal{L}(\mathcal{H})$, then for every $\epsilon > 0$, there exists a diagonable operator D_ϵ in $\mathcal{L}(\mathcal{H})$ such that $N \sim D_\epsilon(\epsilon)$. Furthermore, if N_1 and N_2 are normal operators in $\mathcal{L}(\mathcal{H})$, then $N_1 \sim N_2$ if and only if $\sigma_e(N_1) = \sigma_e(N_2)$. Moreover, if $\sigma(N_1) = \sigma_e(N_1) = \sigma_e(N_2) = \sigma(N_2)$, then $N_1 \sim N_2(\epsilon)$ for every positive number ϵ.

COROLLARY 1. If N_1 and N_2 are normal operators in $\mathcal{L}(\mathcal{H})$, then $N_2 \sim N_1 \oplus N_2$ if and only if $\sigma_e(N_1) \subset \sigma_e(N_2)$.

THEOREM 2 (Pearcy-Salinas [7], Salinas [12]). Let N and T be, respectively, a normal and an essentially hyponormal operator in $\mathcal{L}(\mathcal{H})$. Then $T \sim N \oplus T$ if and only if $\sigma_e(N) \subset \sigma_{\ell e}(T)$. Moreover, if $\sigma(N) \subset \sigma_{\ell e}(T)$, then for every $\epsilon > 0$, $T \sim N \oplus T(\epsilon)$.

Before stating the third and most impressive theorem of this sequence, we need another definition.

DEFINITION 2 ([6]). If $T \in \mathcal{L}(\mathcal{H})$, a <u>hole</u> in $\sigma_e(T)$ is a bounded component of $\mathbb{C}\setminus\sigma_e(T)$, and a <u>pseudohole</u> in $\sigma_e(T)$ is a component of $\sigma_e(T)\setminus\sigma_{\ell e}(T)$ $[\sigma_e(T)\setminus\sigma_{re}(T)]$. (One knows from the Fredholm theory that every hole H in $\sigma_e(T)$ is associated with a unique integer, namely the Fredholm index of $T - \lambda$ for every λ in H, and one also knows that every pseudohole pH in $\sigma_e(T)$ is open in $\mathbb{C}$ and is associated with a unique number $+\infty$ or $-\infty$, namely the Fredholm index of $T - \lambda$ for every λ in pH.) The <u>spectral picture</u> of an operator T (notation: $SP(T)$) is the structure consisting of the set $\sigma_e(T)$, the collection of holes and pseudoholes in $\sigma_e(T)$, and the corresponding set of Fredholm indices.

It is elementary that if $T_1, T_2 \in \mathcal{L}(\mathcal{H})$ and $T_1 \sim T_2$, then $SP(T_1) = SP(T_2)$. In other words, the spectral picture of an operator is an invariant for the relation of compalence.

THEOREM 3 (Brown-Douglas-Fillmore [4]). If T_1 and T_2 are essentially normal operators in $\mathcal{L}(\mathcal{H})$, then $T_1 \sim T_2$ if and only if $SP(T_1) = SP(T_2)$.

The reader will observe that the above theorems all concern the relation of compalence and the classes of normal and essentially normal operators. The papers [10] and [11] were written because the authors perceived that there was a good possibility that Theorems 1, 2, and 3 could be generalized so as to pertain to a larger class of operators that was close at hand.

DEFINITION 3. If $T \in \mathcal{L}(\mathcal{H})$, we denote the C^*-algebra generated by T and $1_{\mathcal{H}}$ by $C^*(T)$, and the C^*-algebra $\pi(C^*(T))$ (which is the C^*-algebra generated by $\pi(T)$ and 1) by $C_e^*(T)$. An operator T in $\mathcal{L}(\mathcal{H})$ is <u>algebraically n-normal</u> (where n is a positive integer) if the algebra $C^*(T)$ satisfies the identity

$$\sum_{\sigma} (\operatorname{sgn} \sigma) X_{\sigma(1)} X_{\sigma(2)} \cdots X_{\sigma(2n)} = 0 \tag{1}$$

where the index σ runs over all permutations on $2n$ objects. (To say that $C^*(T)$ satisfies this identity means that if any $2n$ operators are chosen from $C^*(T)$ and substituted for the $X_{\sigma(i)}$ in (1), then the resulting operator is always zero.) An operator T in $\mathcal{L}(\mathcal{H})$ is <u>n-normal</u> if T is unitarily equivalent to an $n \times n$ matrix $(T_{ij})_{i,j=1}^n$, acting on the direct sum of n copies of $\mathcal{H}$, whose entries T_{ij} are mutually commuting normal operators on $\mathcal{H}$.

The connection between the concepts of algebraically n-normal and n-normal operators is given by the following proposition.

PROPOSITION 1 (Brown [3]). Every n-normal operator is algebraically n-normal. Every algebraically n-normal operator T is a direct sum $T = \sum_{k=1}^{n} \oplus T_k$ where each T_k is k-normal. (Note that zero is an n-normal operator for every positive integer n.)

The classes of n-normal and algebraically n-normal operators have been much studied. (See, for example, the bibliography of [6].) In particular, it is easy to verify that the set of algebraically 1-normal operators on $\mathcal{H}$ coincides with the set of 1-normal operators on $\mathcal{H}$, and both of these sets coincide with the set of normal operators on $\mathcal{H}$.

DEFINITION 4. An operator T in $\mathcal{L}(\mathcal{H})$ is <u>essentially algebraically</u> n-normal if the C^*-algebra $C_e^*(T)$ satisfies the identity (1). Furthermore, T is <u>essentially n-normal</u> if T is unitarily equivalent to an $n \times n$ operator matrix $(T_{ij})_{i,j=1}^n$ acting on the direct sum of n copies of $\mathcal{H}$ and having the property that the set $\{\pi(T_{ij})\}_{i,j=1}^n$ consists of mutually commuting normal elements of $\mathcal{L}(\mathcal{H})/\mathbb{K}(\mathcal{H})$.

The results and techniques of [3] yield certain facts about the relationship between these last two classes of operators.

PROPOSITION 2. Every [algebraically] n-normal operator is essentially [algebraically] n-normal, and every essentially n-normal operator is essentially algebraically n-normal.

Whether every essentially algebraically n-normal operator is a direct sum of essentially n-normal operators is not clear to the authors (cf. [13]), and seems to be a nontrivial problem ([6], Prob. D). In any case, it is easy to see that the three sets of essentially normal operators on $\mathcal{H}$, essentially 1-normal operators on $\mathcal{H}$, and essentially algebraically 1-normal operators on $\mathcal{H}$ all coincide.

The fact that the classes of [algebraically] n-normal and essentially [algebraically] n-normal operators were available as generalizations of the classes of normal and essentially normal operators led the authors to carry out the study [10], [11], [13]. The first problem we faced was what to use for new invariants to take the place of those appearing in Theorems 1, 2, and 3, and in this connection it is important to review now the principal results of [12], since that paper pointed the way.

DEFINITION 5. If $T \in \mathcal{L}(\mathcal{H})$, then the <u>reducing essential spectrum</u> of T (notation: $R_e(T)$) is the set of all complex numbers λ such that $T \sim \lambda_{\mathcal{H}} \oplus T$.

PROPOSITION 3. For every T in $\mathcal{L}(\mathcal{H})$, $R_e(T)$ is identical with the set of all complex numbers λ such that for every positive number ϵ, there exists an operator T_ϵ such that $T \sim \lambda_{\mathcal{H}} \oplus T_\epsilon(J_\epsilon)$. Furthermore, $R_e(T)$ consists exactly of the set of all complex numbers λ such that there exists an orthonormal sequence $\{x_n\}$ of vectors in $\mathcal{H}$ satisfying

$$\|(T-\lambda)x_n\| + \|(T^*-\bar{\lambda})x_n\| \to 0 \tag{2}$$

DEFINITION 6. If $T \in \mathcal{L}(\mathcal{H})$, then the <u>reducing spectrum</u> of T (notation: $R(T)$) consists of the set of all complex numbers λ with the property that there exists a sequence $\{x_n\}$ of unit vectors in $\mathcal{H}$ such that (2) is valid.

THEOREM 4 ([12]). The following statements are valid.

1) For every T in $\mathcal{L}(\mathcal{H})$, $R_e(T)$ is compact and satisfies $R_e(T) \subset \sigma_{\ell e}(T) \cap \sigma_{re}(T)$.

2) There exist operators T in $\mathcal{L}(\mathcal{H})$ for which $R_e(T) = \emptyset$.

3) If $T_1 \sim T_2$, then $R_e(T_1) = R_e(T_2)$.

4) If T is essentially hyponormal, then $R_e(T) = \sigma_{\ell e}(T)$.

5) If T is essentially normal, then $R_e(T) = \sigma_e(T)$.

6) If $N, T \in \mathcal{L}(\mathcal{H})$ and N is normal, then $T \sim N \oplus T$ if and only if $R_e(N) \subset R_e(T)$. Moreover, if $\sigma(N) \subset R_e(T)$, then for every $\epsilon > 0$, $T \sim N \oplus T(\epsilon)$.

7) If $T \in \mathcal{L}(\mathcal{H})$, then $R(T) = R_e(T) \cup R_0(T)$, where $R_0(T)$ is the set of all isolated reducing eigenvalues λ of T whose

associated reducing eigenspace is finite dimensional.

Note that 6) is a generalization of Theorem 2 by virtue of 4) and 5). Moreover, by virtue of 5), Theorems 1 and 3 may be restated in terms of the compalence invariant $R_e(T)$. Thus in trying to generalize Theorems 1, 2, 3, and 4 to the setting of n-normal and essentially n-normal operators, it was natural to ask what invariant would take the place of $R_e(T)$. Fortunately for us, the sets $R(T)$ and $R_e(T)$ were given a different characterization by S. G. Lee in his thesis (U. of Cal. at Santa Barbara).

PROPOSITION 4 (Lee). If T is any operator in $\mathcal{L}(\mathcal{H})$, then $R(T)$ $[R_e(T)]$ consists exactly of those complex numbers λ for which there exists a character $\varphi[\varphi_e]$ on $C^*(T)[C_e^*(T)]$ such that $\varphi(T) = \lambda$ $[\varphi_e(\pi(T)) = \lambda]$. Equivalently, $R_e(T)$ consists of all those complex numbers λ such that there exists a character ψ on $C^*(T)$ such that $\psi(T) = \lambda$ and such that ψ factors through the Calkin algebra.

Since the characters on a C^*-algebra are, by definition, the unital (i.e., identity preserving) 1-dimensional representations of the algebra, this gave us an idea of what to use for an invariant to replace $R_e(T)$ when dealing with n-normal operators.

DEFINITION 7. If $\mathcal{C}$ is a unital C^*-algebra, then an <u>n-dimensional</u> representation of $\mathcal{C}$ is a representation φ of $\mathcal{C}$ into the C^*-algebra $\mathbb{M}_n$ of $n \times n$ complex matrices. If $\varphi(\mathcal{C}) = M_n$, then φ is <u>irreducible</u>, and if $\varphi(1_{\mathcal{C}}) = 1_n$, then φ is <u>non-degenerate</u>.

To see whether the non-degenerate n-dimensional representations of $C^*(T)$ and $C_e^*(T)$ might be pressed into service as useful invariants, it was first necessary to answer the question: Can one characterize the (irreducible) n-dimensional representations of the C^*-algebra $C^*(T)$ for an operator T in $\mathcal{L}(\mathcal{H})$? A little thought shows that there are least two ways that an irreducible n-dimensional representation of $C^*(T)$ can arise:

a) Suppose that K is an irreducible matrix in $\mathbb{M}_n$ and that T is unitarily equivalent to an operator on $\mathcal{H}$ having a matrix of the form

$$\left(\begin{array}{cc|c} K & & 0 \\ & \ddots & \\ 0 & & K \\ \hline & 0 & T_1 \end{array}\right)$$

where there are p copies of the matrix K on the diagonal (for some positive integer p). In this situation, it is clear that if one defines $\varphi(T) = K$ and $\varphi(1_{\mathcal{H}}) = 1_n$, then φ can be extended to an irreducible n-dimensional representation of $C^*(T)$.

b) With K as before, suppose that $\pi(T)$ has the form

$$\pi(T) = \left(\begin{array}{c|c} K & 0 \\ \hline 0 & T_1 \end{array}\right)$$

in the Calkin algebra $\mathcal{L}(\mathcal{H})/\mathbb{K}(\mathcal{H})$. (To be precise, this means that there exist $n+1$ equivalent, orthogonal, nonzero projections E_{ii} , $1 \leq i \leq n+1$, in the Calkin algebra with sum 1 and with implementing partial isometries E_{ij} , $1 \leq i,j \leq n+1$, $i \neq j$ (i.e., the set $\{E_{ij}\}_{i,j=1}^{n+1}$ is a system of matrix units), with the following properties: $\pi(T)E_{n+1,n+1} = E_{n+1,n+1}\pi(T)$ and $\pi(T)(1 - E_{n+1,n+1}) = \sum_{i,j=1}^{n} \kappa_{ij}E_{ij}$ where K is the matrix $[\kappa_{ij}]$.) Then if one defines $\varphi_e(\pi(T)) = K$ and $\varphi_e(1) = 1_n$, it is clear that φ_e can be extended to an irreducible n-dimensional representation of $C_e^*(T)$, and thus that $\psi = \varphi_e \circ \pi$ is an irreducible n-dimensional representation of $C^*(T)$.

The remarkable fact is that these two ways are the only ways that irreducible n-dimensional representations of $C^*(T)$ can arise (up to unitary equivalence).

THEOREM 5 ([10]). Let T belong to $\mathcal{L}(\mathcal{H})$, and let φ be an irreducible n-dimensional representation of $C^*(T)$. Then either

a) $C^*(T) \cap \mathbb{K}(\mathcal{H}) \not\subset \ker \varphi$, in which case there exists a projection Q in $C^*(T)$ that commutes with $C^*(T)$ and has rank an integral multiple of n such that $\varphi(Q) = 1$ and such that $\varphi|C^*(T)Q$ is a C^*-algebra isomorphism of the subalgebra $C^*(T)Q$ onto $\mathbb{M}_n$, or

b) $C^*(T) \cap \mathbb{K}(\mathcal{H}_T) \subset \ker \varphi$, in which case there exists a pair (P, η) where P is a projection in $\mathcal{L}(\mathcal{H})$ of infinite rank and nullity such that $\pi(P)$ commutes with $C_e^*(T)$ and η is a C^*-algebra isomorphism of $C_e^*(T)\pi(P)$ onto $\mathbb{M}_n$ such that $\varphi(A) = \eta(\pi(A)\pi(P))$ for every A in $C^*(T)$.

We have stated this result for singly generated C^*-subalgebras for simplicity. Actually, the result is valid for all separably acting separable C^*-algebras, and it is stated and proved in that generality in [10].

Using Theorem 5 it is possible to describe all non-degenerate n-dimensional representations of the C^*-algebra $C^*(T)$ (and, more generally, all non-degenerate n-dimensional representations of any separably acting separable C^*-algebra; see [10]). We now outline this description. Let $T \in \mathcal{L}(\mathcal{H})$, and let φ' be a non-degenerate n-dimensional representation of $C^*(T)$. Then there exists a representation φ of $C^*(T)$ that is unitarily equivalent to φ' and a set $\{\varphi_{n_1}, \ldots, \varphi_{n_m}\}$ of irreducible, non-equivalent, n_i-dimensional subrepresentations of φ such that

$$\varphi(A) = \sum_{k=1}^{m} \oplus (\varphi_{n_k}(A) \otimes 1_{j_k}) \quad , \quad A \in C^*(T) .$$

(The numbers j_k are the multiplicities of the representations φ_{n_k}.) The set $\{\varphi_{n_1}, \ldots, \varphi_{n_m}\}$ can be partitioned into two subsets

Γ_e and Γ_ι consisting, respectively, of those φ_{n_k} that are essential (i.e., that factor through the Calkin algebra) and those that are inessential (i.e., those that do not so factor). If one defines

$$\varphi_e = \sum_{k \in \Gamma_e} \oplus (\varphi_{n_k} \otimes 1_{j_k}), \varphi_\iota = \sum_{k \in \Gamma_\iota} \oplus (\varphi_{n_k} \otimes 1_{j_k}),$$

Then $\varphi = \varphi_e \oplus \varphi_\iota$ and 1_n can be written as a direct sum $1_n = E_e \oplus E_\iota$ where E_e is the projection in $\mathbb{M}_n$ that is the identity of the algebra $\varphi_e(C^*(T)) \oplus 0$ and similarly for E_ι.

THEOREM 6 ([10]). With the notation as established above, there exists a projection Q of finite rank in $C^*(T)$ such that Q commutes with $C^*(T)$, $\varphi(Q) = E_\iota$, $Q \neq 0$ if and only if $E_\iota \neq 0$, and $\varphi_\iota | \mathfrak{A}Q$ is a C^*-algebra isomorphism. Moreover, if $C^*(T)(1-Q) \neq 0$, then $\varphi | C^*(T)(1-Q)$ factors through the Calkin algebra, and $\varphi | C^*(T)(1-Q)$ is nontrivial if and only if φ_e is. In this case, there exists a pair (P, η) where P is a projection in $\mathcal{L}(\mathcal{H})$ of infinite rank and nullity such that $\pi(P)$ commutes with $\pi(C^*(T))$ and η is a C^*-algebra isomorphism of $\pi(C^*(T))$ into $\mathbb{M}_n$ such that $\varphi_e(A) = \eta(\pi(A)\pi(P))$ for all A in $C^*(T)$.

This successful characterization of the non-degenerate n-dimensional representations of $C^*(T)$ for T in $\mathcal{L}(\mathcal{H})$ made it worthwhile to give the appropriate definitions which generalize the notions of the reducing spectrum and the reducing essential spectrum of an operator.

DEFINITION 8 ([10]). If T belongs to $\mathcal{L}(\mathcal{H})$ and n is any positive integer, then the reducing $n \times n$ spectrum of T (notation: $R^n(T)$) is the set of all L in $\mathbb{M}_n$ such that there exists a non-degenerate (but not necessarily irreducible) n-dimensional representation φ of $C^*(T)$ such that $\varphi(T) = L$. The reducing essential $n \times n$ spectrum of T (notation: $R^n_e(T)$) is the set of all L in

$\mathbb{M}_n$ for which there exists a non-degenerate, n-dimensional representation φ_e of $C_e^*(T)$ such that $\varphi_e(\pi(T)) = L$.

We summarize in the following theorem some of the properties of the sets $R^n(T)$ and $R_e^n(T)$.

THEOREM 7 ([10],[11]). If T belongs to $\mathcal{L}(\mathcal{H})$ and n is any positive integer, then the following statements are valid:

1) $R^1(T) = R(T)$ and $R_e^1(T) = R_e(T)$.

2) $R_e^n(T) \subset R^n(T)$.

3) There exists a T_0 in $\mathcal{L}(\mathcal{H})$ for which all of the sets $R^n(T_0)$ are void.

4) $R^n(T)[R_e^n(T)]$ is a union of unitary equivalence classes in $\mathbb{M}_n$.

5) $R^n(T)[R_e^n(T)]$ is a unitary invariant for the operator T .

6) If $R^n(T)[R_e^n(T)]$ is nonvoid, then so is $R^{kn}(T)[R_e^{kn}(T)]$ for every positive integer k .

7) $R^n(T)[R_e^n(T)]$ is compact.

8) If $T \sim T'$, then $R_e^n(T) = R_e^n(T')$.

The following theorem gives a spatial characterization of the sets $R_e^n(T)$ that is the analog of Proposition 3, which is concerned with $R_e^1(T) = R_e(T)$.

THEOREM 8 ([10]). If $T \in \mathcal{L}(\mathcal{H})$, n is any positive integer, and $L \in \mathbb{M}_n$, then the following statements are equivalent:

a) $L \in R_e^n(T)$.

b) There exists a sequence $\{B_j\}_{j=1}^{\infty}$ of isometries $B_j : \mathbb{C}_n \to \mathcal{H}$ with mutually orthogonal ranges such that

$$\|TB_jB_j^* - B_jLB_j^*\| + \|T^*B_jB_j^* - B_jL^*B_j^*\| \to 0 . \tag{3}$$

c) There exists a sequence $\{B_j\}$ of isometries $B_j : \mathbb{C}_n \to \mathcal{H}$ that converges weakly to zero and satisfies (3).

d) For every positive number ϵ , there exists a system

$\{W_{i,j}\}_{i,j=1}^{n}$ of matrix units in $\mathcal{L}(\mathcal{H})$ satisfying

i) The projection $Q = \sum_{i=1}^{n} W_{i,i}$ has infinite rank and nullity.

ii) $QT - TQ \in J(\mathcal{H})$ and $\|QT - TQ\|_J < \epsilon$.

iii) $W_{i,i} T W_{j,j} - [L]_{i,j} W_{i,j} \in J(\mathcal{H})$ and has trace-norm less than ϵ, $1 \leq i,j \leq n$, where $[L]_{i,j}$ is the (i,j) entry of L.

Clearly condition d) says that if $L \in R_e^n(T)$, then there is a subspace $\mathcal{M}$ of the space $\mathcal{H}$ such that $\mathcal{M}$ reduces T up to a trace-class perturbation of small trace-norm and such that $T|\mathcal{M}$ acts like the $n \times n$ matrix L up to a similar perturbation.

The following result concerning the upper semi-continuity of the sets $R^n(T)$ and $R_e^n(T)$ proved to be important.

PROPOSITION 5 ([11]). Let T belong to $\mathcal{L}(\mathcal{H})$, n be a positive integer, and Ω be an open set in M_n such that $\Omega \supset R^n(T)$ $[\Omega \supset R_e^n(T)]$. Then there exists an $\epsilon > 0$ such that for every S in $\mathcal{L}(\mathcal{H})$ with $\|S - T\| < \epsilon$, one has $R^n(S) \subset \Omega [R_e^n(S) \subset \Omega]$. In particular, if $R^n(T)$ $[R_e^n(T)]$ is void, then $R^n(S) [R_e^n(S)]$ is void for all S sufficiently close to T.

Using these facts and techniques from [7] and [12], the authors were able to prove in [11] the desired generalizations of Theorems 1,2 and 4. The following preliminary result, which is generalization of Berg's theorem [2] was obtained first in [4]. Recall that an n-normal operator in $\mathcal{L}(\mathcal{H})$ is said to be diagonable if it is unitarily equivalent to an $n \times n$ matrix (S_{ij}) where the S_{ij} are simultaneously diagonable normal operators on $\mathcal{H}$.

THEOREM 9 ([4], [11]). Let T be an n-normal operator. Then $R_e^n(T)$ is nonempty, and for every $\epsilon > 0$, there exists a diagonable n-normal operator S_ϵ such that $T \sim S_\epsilon(\epsilon)$.

Here are the promised generalizations of Theorems 2 and 1 to the setting of n-normal operators.

THEOREM 10 ([11]). Suppose that S and T are operators in $\mathcal{L}(\mathcal{H})$

and that S is n-normal. Then $T \sim S \oplus T$ if and only if $R_e^n(S) \subset R_e^n(T)$. Furthermore, if $R^n(S) \subset R_e^n(T)$, then $T \sim S \oplus T(\epsilon)$ for every positive number ϵ.

THEOREM 11 ([11]). Suppose that S and T are n-normal operators in $\mathcal{L}(\mathcal{H})$. Then $S \sim T$ if and only if $R_e^n(S) = R_e^n(T)$. Furthermore, if $R^n(S) = R_e^n(S) = R_e^n(T) = R^n(T)$, then $S \sim T(\epsilon)$ for every positive number ϵ.

Proof. If $S \sim T$, then $R_e^n(S) = R_e^n(T)$ by Theorem 7. If $R_e^n(S) = R_e^n(T)$, then $S \sim S \oplus T \sim T$ by Theorem 10, and if, in addition, $R^n(S) = R_e^n(S) = R_e^n(T) = R^n(T)$, then for every $\epsilon > 0$, $S \sim S \oplus T(\epsilon/2)$ and $T \sim S \oplus T(\epsilon/2)$, so $S \sim T(\epsilon)$.

We turn our attention now to the problem of generalizing Theorem 3. The class of operators with which we shall be concerned is the class of essentially n-normal operators in $\mathcal{L}(\mathcal{H})$ (Definition 4). As we saw in Theorem 7, if S and T belong to this class of operators and $S \sim T$, then $R_e^n(S) = R_e^n(T)$. Thus it would appear that what is needed is an additional invariant I_T to combine with the set $R_e^n(T)$ and having the property that if S and T are essentially n-normal and $R_e^n(S) = R_e^n(T)$, $I_S = I_T$, then $S \sim T$. (In the case $n = 1$ this invariant is the set of integers associated with the holes in $\sigma_e(T) = R_e(T)$.) But the situation is more complicated when $n > 1$, as the following considerations show.

DEFINITION 9 ([4], [6]). If $\mathcal{H}_1$ and $\mathcal{H}_2$ are Hilbert spaces of dimension $\aleph_0$ and $V: \mathcal{H}_1 \to \mathcal{H}_2$ is an isometry [co-isometry], then V is said to have <u>finite multiplicity</u> if the projection $1_{\mathcal{H}_2} - VV^*$ [$1_{\mathcal{H}_1} - V^*V$] has finite rank. If S and T act on $\mathcal{H}_1$ and $\mathcal{H}_2$, respectively, and there exists an isometry or co-isometry of finite multiplicity $V: \mathcal{H}_1 \to \mathcal{H}_2$ such that $VSV^* + K = T$ for some K in $\mathbb{K}(\mathcal{H}_2)$, then S and T are <u>weakly compalent</u> (notation: $S \overset{w}{\sim} T$).

PROPOSITION 6. Weak compalence is an equivalence relation.

Compalent operators are weakly compalent, and essentially normal operators that are weakly compalent are also compalent.

Proof. Since the product of two isometries [co-isometries] of finite multiplicity is again an isometry [co-isometry] of finite multiplicity, to prove the first statement, it suffices to prove that the product of an isometry of finite multiplicity and a co-isometry of finite multiplicity (in either order) differs by a compact operator from some third isometry or co-isometry of finite multiplicity, and this follows, after applying some preliminary unitary equivalence to move all of the operators to the same space, from the fact that all unitary operators in a Calkin algebra can be lifted to such an isometry or co-isometry ([4, Theorem 3.1]). The second statement is obvious, and the third is Theorem 4.3 of [4].

PROPOSITION 7. If S and T belong to $\mathcal{L}(\mathcal{H})$ and $S \overset{w}{\sim} T$, then $R_e^n(S) = R_e^n(T)$ for every positive integer n.

Proof. This follows immediately from the fact that the C^*-algebras $C_e^*(S)$ and $C_e^*(T)$ are isomorphic.

We saw in Proposition 6 that on the class of essentially normal operators, compalence and weak compalence are the same equivalence relation. The following example, which has been around for some time, shows that on the class of essentially 2-normal operators, this is not the case.

EXAMPLE A. Let U be a unilateral shift operator of multiplicity one in $\mathcal{L}(\mathcal{H})$, and let S be the operator in $\mathcal{L}(\mathcal{H}\oplus\mathcal{H})$ given by the matrix

$$\begin{pmatrix} 0 & U \\ 0 & 0 \end{pmatrix}.$$

Let T be the operator in $\mathcal{L}(\mathcal{H}\oplus\mathcal{H})$ given by the matrix

$$\begin{pmatrix} 0 & 1 \\ 0 & 0 \end{pmatrix},$$

and observe that $S \overset{w}{\sim} T$ by virtue of the following calculation:

$$\begin{pmatrix} U^* & 0 \\ 0 & 1 \end{pmatrix} \cdot \begin{pmatrix} 0 & U \\ 0 & 0 \end{pmatrix} \cdot \begin{pmatrix} U & 0 \\ 0 & 1 \end{pmatrix} = \begin{pmatrix} 0 & 1 \\ 0 & 0 \end{pmatrix}.$$

Suppose now that $S \sim T$, and let $W = (W_{ij})$ and $K = (K_{ij})$ be a unitary operator and a compact operator, respectively, in $\mathcal{L}(\mathcal{U}\oplus\mathcal{U})$ such that $WTW^* + K = S$. Matricial calculation shows that the operators $W_{11}W_{12}^*$, $W_{21}W_{12}^*$, and $W_{21}W_{12}^*$, and $W_{21}W_{22}^*$ are compact, and since $W_{12}^*W_{12}+W_{22}^*W_{22} = 1$, we see that W_{21} is compact. Since also $W_{11}^*W_{11} + W_{21}^*W_{21} = 1$, it follows that $W_{11}^*W_{11} = 1 + K'$ where K' is compact. Hence $W_{11}^*W_{11}W_{12}^* = (1 + K)W_{12}^*$ is compact, from which it follows that W_{12} is compact. This implies that $W = (W_{11} \oplus W_{22}) + K''$ where K'' is compact, and since W is unitary, it follows that W_{11}, W_{22} are Fredholm operators and that $i(W_{11}) + i(W_{22}) = 0$. On the other hand, since $W_{11}W_{22}^* = U + K_{12}$, we have $i(W_{11}) + i(W_{22}^*) = -1$, and a subtraction yields $i(W_{22}) - i(W_{22}^*) -1$, a manifest impossibility. Thus S and T are not compalent, and we can further deduce that S is not compalent to any 2-normal operator B as follows. Suppose $S \sim B$. Then $B \sim S \overset{w}{\sim} T$, and hence by Proposition 7, $R_e^2(B) = R_e^2(T)$. Thus by Theorem 11, $B \sim T$. But then $S \sim B \sim T$, and we just saw that $S \sim T$ is false.

This example shows that the proper relation to study to generalize Theorem 3 to the setting of essentially n-normal operators is not compalence but <u>weak</u> compalence. The appropriate definitions to begin with are the following.

DEFINITION 10 ([13]). Let n be any positive integer, and let X be a nonempty compact subset of $\mathbf{M}_n$ that is a union of unitary equivalence classes in $\mathbf{M}_n$ and has the further property that if j is any positive integer and $L_i \oplus M_i \in X$, $1 \leq i \leq j$, where $L_i \in \mathbb{M}_{k_i}$ and $\sum_{i=1}^{j} k_i = n$, then also $\sum_{i=1}^{j} \oplus L_i \in X$. Then **X** is said to be a <u>hypoconvex</u> subset of $\mathbb{M}_n$.

DEFINITION 11 ([6], [13]). Let n be any positive integer, and let X be a hypoconvex subset of $\mathbb{M}_n$. We denote by $EN_n(X)$ the set of all essentially n-normal operators T in $\mathcal{L}(\mathcal{H})$ such that $R_e^n(T) = X$. According to Proposition 6, weak compalence is an equivalence relation on $EN_n(X)$, and we define Ext(X) to be the collection of equivalence classes into which weak compalence partitions $EN_n(X)$, i.e., $\text{Ext}(X) = EN_n(X)/\overset{w}{\sim}$.

Observe that when $n = 1$, then X is an arbitrary nonempty compact subset of $\mathbb{C}$, and $EN_1(X)$ is the set of all essentially normal operators T in $\mathcal{L}(\mathcal{H})$ such that $\sigma_e(T) = X$ (Theorem 4). Since compalence and weak compalence are the same equivalence relation on this set (Proposition 6), Ext (X) as defined above coincides with the set Ext(X) as defined in [1] and [4].

The next step is to show that Ext(X), as given by Definition 11, can be turned into a group. In this connection, we first fix a Hilbert space isomorphism ρ of $\mathcal{H}$ onto $\mathcal{H}\oplus\mathcal{H}$. A calculation shows that if A and B are essentailly n-normal operators in $\mathcal{L}(\mathcal{H})$, then $\rho^{-1}(A\oplus B)\rho$ is also an essentially n-normal operator in $\mathcal{L}(\mathcal{H})$. Moreover, if A' and B' are essentially n-normal operators in $\mathcal{L}(\mathcal{H})$ such that $A \overset{w}{\sim} A'$ and $B \overset{w}{\sim} B'$, then it is easy to verify that $\rho^{-1}(A'\oplus B')\rho \overset{w}{\sim} \rho^{-1}(A\oplus B)\rho$. Furthermore, the second author has recently shown (cf. [13]) that if $R_e^n(A) = R_e^n(B) = X$, then $R_e^n(\rho^{-1}(A \oplus B)\rho) = X$ also. These facts show that the following definition can be made.

DEFINITION 12. If n is any positive integer and X is a given hypoconvex subset of $\mathbb{M}_n$, then the operation of addition on Ext(X) is defined by

$$[A] + [B] = [\rho^{-1}(A \oplus B)\rho], \; A,B \in EN_n(X).$$

It is routine to verify the following proposition.

PROPOSITION 8. Addition is an associative and commutative binary operation on Ext (X).

To turn Ext(X) into an abelian semigroup an identity is needed, and the following result provides one.

PROPOSITION 9 ([13]). There exists an n-normal operator N in $\mathcal{L}(\mathcal{H})$ such that [N] $\in$ Ext(X), i.e., such that $R_e^n(N) = X$.

THEOREM 12. If N is as in Proposition 9, then [N] is an identity for Ext(X), i.e., [N] + [A] = [A] for every [A] in Ext (X).

Proof. Since N is n-normal and $R_e^n(N) = R_e^n(A)$, it follows from Theorem 10 that $N \oplus A \sim A$, and it results easily that $\rho^{-1}(N \oplus A)\rho \overset{w}{\sim} A$, as desired.

To complete the argument that Ext(X) is a group, it must be shown that if [T] $\in$ Ext(X) and [N] is as above, then there exists an essentially n-normal operator R in $EN_n(X)$ such that [R] + [T] = [N], and this was recently done by the second author.

THEOREM 13 ([13]). If n is any positive integer and X is any hypoconvex subset of M_n, then Ext (X) forms a group under the operation of addition given in Definition 12 and with the identity given by Theorem 12.

To classify essentially n-normal operators up to weak compalence, and thereby to generalize Theorem 3, it thus suffices to construct an isomorphism of the group Ext(X) (for a given positive integer n and a given hypoconvex subset X of M_n) into a known group G, and this problem remains under study (cf. [13]).

REFERENCES

1. W. Arveson, A note on essentially normal operators, Proc. Royal Irish Acad. Sec. A, 74(1974), 143-146.

2. I.D. Berg, An extension of the Weyl-von Neumann theorem to normal operators, Trans. Amer. Math. Soc. 160(1971), 365-371.

3. A. Brown, Unitary equivalence of binormal operators, Amer. J. Math. 76(1954), 414-434.

4. L. Brown, R.G. Douglas, and P. Fillmore, Unitary equivalence modulo the compact operators and extensions of C*-algebras, Proceedings of a Conference on Operator Theory, Lecture Notes in Mathematics No. 345, Springer Verlag, New York, 1973, 58-128.

5. P.R. Halmos, Limits of shifts, Acta. Sci. Math. (Szeged), 34(1973), 131-139.

6. C. Pearcy, Some recent advances in operator theory, CBMS-NSF regional conference notes, to appear.

7. C. Pearcy and N. Salinas, Compact perturbations of seminormal operators, Indiana Univ. Math. J. 22(1973), 789-793.

8. _________, Operators with compact self-commutator, Canadian J. Math. 26(1974), 115-120.

9. _________, Finite-dimensional representations of separable C*-algebras, Bull. Amer. Math. Soc. 80(1974), 970-972.

10. _________, Finite-dimensional representations of C*-algebras and the reducing matricial spectra of an operator, Revue. Roum. Math. Pures et Appl. 20(1975), 567-598.

11. _________, The reducing essential matricial spectra of an operator, Duke Math. J. 42(1975), 423-434.

12. N. Salinas, Reducing essential eigenvalues, Duke Math. J. 40(1973), 561-580.

13. _________, Extensions of C*-algebras and essentially n-normal operators, Bull. Amer. Math. Soc. 82(1976), 143-146.

K-HOMOLOGY THEORY AND ALGEBRAIC K-THEORY

Graeme Segal, St. Catherine's College, Oxford, England

This talk consists of three parts which are not very closely connected but are all expositions of ideas of Quillen. The first describes classical K-homology theory in a form which may be helpful for understanding the work of Brown, Douglas and Fillmore. The second part is an attempt to explain Quillen's definition of algebraic K-theory as simply as possible and from the point of view closest to classical K-theory. The third part is about classifying spaces, and describes the relationship between the abstract categorical constructions and some more concrete models such as Grassmannians and spaces of Fredholm operators.

1. K Homology Theory

If X is a compact space with a base-point x_0 let $C(X)$ denote the algebra of continuous real-valued functions on X which vanish at x_0. It is a Banach algebra without a 1. We shall regard it as a $*$-algebra with $* = \text{identity}$.

We define a kind of non-abelian spectrum $F(X)$ of $C(X)$ by

$$F(X) = \bigcup_{n \geq 0} \operatorname{Hom}(C(X); M_n)$$

where M_n is the $n \times n$ matrices over $\mathbb{R}$, Hom means $*$-homomorphisms, and the union is taken by embedding M_n in M_{n+1} by $A \longmapsto A \oplus 0$. Each space of homomorphisms is given the compact-open or weak topology--the two coincide.

By the spectral theorem to give a point of $F(X)$ is to give a finite subset S of X not containing x_0, and for each $x \in S$ a finite dimensional subspace V_x of $\mathbb{R}^\infty$ such that V_x and $V_{x'}$ are perpendicular if $x \neq x'$. The topology of $F(X)$ is such that
(a) distinct points x and x' in S can move into coincidence at x'', and then $V_{x''}$ will be the limit of $V_x \oplus V_{x'}$, and (b) a point

$x \in S$ can move to x_0 and is then discarded. If X is path-connected so is $F(X)$.

There is a map $F(X) \to P(X)$, where $P(X)$ is the infinite symmetric product of X, which takes $\{V_x\}_{x \in S}$ to $\sum_{x \in S} (\dim V_x).x$. Let us recall the theorem of Dold-Thom [6] that if X is connected and of the homotopy type of a CW complex then the homotopy group $\pi_i P(X)$ can be identified with $\tilde{H}_i(X)$, the reduced homology of X with intogral coefficients.

<u>PROPOSITION (1.1)</u>. If X is connected $\pi_i F(X) = \widetilde{kO}_i(X)$, and $\pi_i F(X) \to \pi_i P(X) \cong \tilde{H}_i(X)$ is the natural map.

Here kO_* is the connective homology theory associated to periodic real K-theory KO_*. That means that there is a transformation $kO_* \to KO_*$ such that $kO_q(\text{point}) \xrightarrow{\cong} KO_q(\text{point})$ when $q \geq 0$, and $kO_q(X) = 0$ for all X when $q < 0$. These properties characterize kO_* in terms of KO_* up to non-canonical isomorphism; but KO_* is fully determined by kO_* because $KO_q(X)$ is the direct limit of the sequence $kO_q(X) \to kO_{q+8}(X) \to kO_{q+16}(X) \to \ldots$, where the maps are 'Bott periodicity'.

If X is not connected $\pi_0 F(X)$ and $\pi_0 P(X)$ are both the free abelian semigroup on the pointed set $\pi_0 X$, i.e. they are the <u>positive</u> elements of $\tilde{H}_0(X)$. Then to make Proposition (1.1) valid one must replace $F(X)$ by $F_\infty(X) = \tilde{H}_0(X) \times \varinjlim F_c(X)$, where c runs through $\pi_0 F(X)$, and $F_c(X)$ is the corresponding component of $F(X)$. For example $F(S^0) = \coprod_{m \geq 0} BO_m$, the Grassmannian of finite dimensional subspaces of $\mathbb{R}^\infty$, and $F_\infty(S^0) = Z \times BO$.

It makes little difference if one replaces the finite-dimensional matrices M_n by the ring $\mathcal{K}$ of compact operators in a real Hilbert space $\mathcal{H}$. In fact if we define $\hat{F}(X) = \mathrm{Hom}(C(X);\mathcal{K})$ then

<u>PROPOSITION (1.2)</u>. The inclusion $F(X) \to \hat{F}(X)$ is a homotopy equivalence if X is locally contractible at x_0.

PROOF: A point of $\hat{F}(X)$ can be represented $\{V_x\}_{x \in S}$, where S is a subset of X such that any neighbourhood of x_0 contains all but finitely many points of S, and the V_x are mutually orthogonal subspaces of $\mathcal{H}$. By the local contractibility of X at x_0 one can retract $\hat{F}(X)$ on to the subspace with S finite. But the Grassmannians of $\mathbb{R}^\infty$ and $\mathcal{H}$ are homotopy-equivalent, so this subspace is equivalent to $F(X)$.

Let us make Proposition (1.1) more explicit in the case when X is the standard sphere S^q. Let $F_q(n)$ be the set of sequences $A_0, \ldots, A_q$ of commuting symmetric $n \times n$ matrices such that $A_0^2 + \ldots + A_q^2 = 1$. Embed $F_q(n)$ in $F_q(n+1)$ by $(A_0, \ldots, A_q) \longmapsto (A_0 \oplus 1, A_1 \oplus 0, \ldots, A_q \oplus 0)$, and let $F_q = \bigcup_{n \geq 1} F_q(n)$. Then because $\mathbb{R}[t_0, \ldots, t_q]/(t_0^2 + \ldots + t_q^2 - 1)$ is a dense subalgebra of $1 + C(S^q)$ there is a homeomorphism $F_q \cong F(S^q)$. (One takes $(1,0\ldots0)$ as base-point in S^q). We shall prove Proposition (1.1) simultaneously with

PROPOSITION (1.3). F_q is a representing space for kO^q if $q > 0$, i.e. it is a $(q-1)$-fold connected covering space of a representing space for KO^q.

The representing spaces for KO^q are well-known in view of Bott's periodicity theorem. They are periodic with period 8 in q. One convenient model for them can be constructed using Clifford algebras as follows. Let C_q be the real Clifford algebra with generators $e_1, \ldots, e_q$ such that $e_ie_j + e_je_i = -2\delta_{ij}$. Let $\Phi_q(n)$ be the space of symmetric unitary $n \times n$ matrices over C_q. ('Unitary' refers to the automorphism of C_q induced by $e_i \longmapsto -e_i$.) Then Bott's theorem can be formulated as

PROPOSITION (1.4). $\Phi_q = \bigcup_{n \geq 1} \Phi_q(n)$ is a representing space for KO^q if $q > 0$.

The space F_q maps to Φ_q by $(A_0, \ldots, A_q) \longrightarrow A_0 + A_1e_1 + \ldots + A_qe_q$.

This map gives us a map $\pi_i F(X) \to KO_i(X)$ when X has the homotopy type of a finite complex. For let Y be a q-dual of X, for example a compact deformation-retract of the complement of X embedded in $\mathbb{R}^{q+1}$, so that there is a map $Y \wedge X \to S^q$. This induces $Y \wedge F(X) \to F(S^q) = F_q$, and hence $F(X) \to \mathrm{Map}(Y; F_q) \to \mathrm{Map}(Y; \Phi_q)$. But $\pi_i \mathrm{Map}(Y; \Phi_q) = [S^i Y; \Phi_q] = \widetilde{KO}^q(S^i Y) = \widetilde{KO}_i(X)$, so we have $\pi_i F(X) \to \widetilde{KO}_i(X)$. To prove Proposition (1.1) it will suffice to show

(a) $X \longmapsto \pi_* F(X)$ is a homology theory on connected spaces, and

(b) $\pi_i F(S^1) \xrightarrow{\cong} KO_i(S^1)$ for $i \geq 0$.

The second statement is clear because $F_i \to \Phi_i$ is an isomorphism. The first is proved in exact analogy with Dold-Thom's proof [6] of the corresponding statement for $\pi_* P(X)$, i.e. one proves

<u>PROPOSITION (1.5)</u>. If Y is a path connected closed subspace of X and is a neighbourhood deformation retract in X then $F(X) \to F(X/Y)$ is a quasifibration with all its fibres homeomorphic to $F(Y)$.

<u>PROOF</u>: The fibre of $p : F(X) \to F(X/Y)$ at $\{V_x\}_{x \in S}$, where $S \subset X - Y$, is $F(Y; V_S^{\perp})$, the space of families $\{V_y\}_{y \in T}$, where T is a finite subset of $Y - y_0$, the $\{V_y\}_{y \in T}$ are mutually perpendicular subspaces of $V_S^{\perp}$, and $V_S = \bigoplus_{x \in S} V_x$. This is a subspace of $F(Y)$ which is homeomorphic to $F(Y)$. If $F(X/Y)$ is filtered by the subspaces $F_n(X/Y)$ consisting of $\{V_x\}_{x \in S}$ with $\Sigma \dim(V_x) \leq n$ then p is a fibration over each step of the filtration. To prove (1.5) one must show that $F_n(X/Y)$ is a deformation retract of a neighbourhood in $F_{n+1}(X/Y)$, and that the retracting homotopy can be covered by one of a neighbourhood of $p^{-1}F_n(X/Y)$ which takes fibres to fibres by homotopy equivalences. Such a deformation is induced by a homotopy $f_t : (X,Y) \to (X,Y)$ such that f_0 is the identity and $f_1^{-1}(Y)$ is a neighbourhood of Y. The maps of fibres induced by f_1 are of the form $F(Y; (V \oplus W)^{\perp}) \to F(Y; W^{\perp})$, and take a family

$\{V_y\}_{y \in T}$ to the same family indexed by $f_1(T)$ together with a new point z such that $V_z \subset V$. These are homotopy equivalences because z can be moved continuously to the base-point, where it disappears, and then the map obtained is homotopic to the inclusion $F(Y;(V \oplus W)^{\perp}) \to F(Y;W^{\perp})$, which is a homotopy equivalence. (Cf. the end of §2 of McDuff's talk in this volume.)

Proposition (1.4) is now proved too, as we have a map $F_q \to \Phi_q$, and (1.1) implies that $\pi_i F_q = 0$ when $i < q$ and $\pi_i F_q \xrightarrow{\cong} \pi_i \Phi_q$ when $i \geq q$.

Relation with the Work of Brown-Douglas-Fillmore.

Let $\mathcal{A}$ be the real Calkin algebra, and define

$$\mathcal{A}(X) = \bigcup_{n \geq 1} \mathrm{Hom}(C(X); M_n \otimes \mathcal{A}).$$

One consequence of the work of Brown-Douglas-Fillmore is that $\pi_0 \mathcal{A}(X) \cong KO_{-1}(X)$. The present work suffices to define maps $\pi_0 \mathcal{A}(X) \leftrightarrows KO_{-1}(X)$ such that the composite $KO_{-1}(X) \to \pi_0 \mathcal{A}(X) \to KO_{-1}(X)$ is the identity. Because $KO_{-1}(X) = \varinjlim kO_{8k-1}(X)$, where the maps are given by Bott periodicity, to define $KO_{-1}(X) \to \pi_0 \mathcal{A}(X)$ one must give compatible maps $\Omega^{8k-1}F(X) \to \mathcal{A}(X)$ for each k. These are induced by the Wiener-Hopf maps $C(S^{8k-1}) \to \mathcal{A}$ of [2] p. 138, for a point of $\Omega^{8k-1}F(X)$ can be regarded as a homomorphism $C(X) \to M_n \otimes C(S^{8k-1})$.

The map $\pi_0 \mathcal{A}(X) \to KO_{-1}(X)$ is defined by analogy with the above method for $F(X)$. Let Y be a q-dual of X. The map $Y \wedge X \to S^q$ induces $Y \wedge C(S^q) \to C(X)$ and hence $\mathcal{A}(X) \to \mathrm{Map}(Y; \mathrm{Hom}(C(S^q); M_n \otimes \mathcal{A}))$. But $\mathrm{Hom}(C(S^q); M_n \otimes \mathcal{A})$ can be identified with the space $F_q(\mathcal{A}; n)$ of sequences $(A_0, \ldots, A_q)$ of $n \times n$ matrices over $\mathcal{A}$ satisfying the same conditions as for F_q. This $F_q(\mathcal{A}; n)$ maps to $\Phi_q(\mathcal{A}; n)$, the space formed like $\Phi_q(n)$ but over the base-ring $\mathcal{A}$ instead of $\mathbb{R}$. It is well-known [3] that $\Phi_q(\mathcal{A}) = \bigcup \Phi_q(\mathcal{A}; n)$ is a representing-space for KO^{q+1}. Thus $\pi_i \mathcal{A}(X) \to \pi_i \mathrm{Map}(Y; \Phi_q(\mathcal{A})) = KO^{q-i+1}(Y) = KO_{i-1}(X)$.

Finally the composition $\widetilde{kO}_{8k-1}(X) \to \pi_0\mathcal{O}(X) \to \widetilde{KO}_{-1}(X)$ is the natural map. That is because Bott periodicity corresponds to a map $\Omega^{8k-1}\Phi_q \to \Phi_q(\mathcal{O})$ of the representing spaces which, when $\Omega^{8k-1}\Phi_q$ is regarded as $\Phi_q(C(S^{8k-1}))$ in the obvious sense, is induced by the Wiener-Hopf map $C(S^{8k-1}) \to \mathcal{O}$. More precisely, we have a commutative diagram

$$\begin{array}{ccc} \Omega^{8k-1}F(X) & \longrightarrow & \mathcal{O}(X) \\ \downarrow & & \downarrow \\ \mathrm{Map}(Y;\Omega^{8k-1}\Phi_q) & \xrightarrow[\text{Bott}]{} & \mathrm{Map}(Y;\Phi_q(\mathcal{O})) \end{array}$$

and we have seen that the left-hand vertical map induces the natural map $\widetilde{kO}_*(X) \to \widetilde{KO}_*(X)$ on homotopy groups.

2. The Definition of Algebraic K-Theory.

One can approach Quillen's definition of algebraic K-theory by analogy with ordinary real or complex K-theory which arises from vector bundles. Thinking of a vector bundle on a space X as a family of vector spaces parametrized by the points of X one tries to construct new cohomology theories by considering families or bundles of finitely generated projective modules over rings other than $\mathbb{R}$ or $\mathbb{C}$.

Recall that the set $\mathrm{Vect}(X)$ of isomorphism classes of (real or complex) vector bundles on a compact space X is an abelian semi-group under the fibrewise direct sum. $K(X)$ is defined as the abelian group formed by adjoining negative elements to $\mathrm{Vect}(X)$. It is a contravariant functor of X because a map $f : X \to Y$ induces a 'pulling-back' operation $f^* : \mathrm{Vect}(Y) \to \mathrm{Vect}(X)$. Two of its basic properties are

(a) it is a homotopy functor, i.e. homotopic maps $X \to Y$ induce the same homomorphism $K(Y) \to K(X)$, and

(b) it has the Mayer-Vietoris property, i.e. whenever X is the union of two compact subspaces Y and Z the sequence

$$K(X) \to K(Y) \oplus K(Z) \to K(Y \cap Z)$$

is exact, where the first map is the sum and the second the difference of the two restrictions.

By Brown's theorem ([5],[1]) properties (a) and (b) together are equivalent to the representability of the functor K, i.e. to the fact that $K(X) \cong [X;B]$ for some space B, where $[X;B]$ denotes the homotopy-classes of maps $X \to B$.

The proof of the Mayer-Vietoris property has two ingredients. Because of the 'clutching' process for vector bundles the sequence

$$\mathrm{Vect}(X) \to \mathrm{Vect}(Y) \times \mathrm{Vect}(Z) \rightrightarrows \mathrm{Vect}(Y \cap Z)$$

is exact, i.e. bundles on Y and Z whose restrictions to $Y \cap Z$ are isomorphic can be joined together to produce a bundle on $X = Y \cup Z$. But 'clutching' does not carry over automatically from the semigroup to the group: one needs the fact that for any bundle E on a compact space there is a 'complementary' bundle $E^{\perp}$ such that $E \oplus E^{\perp}$ is trivial. This means that any element of $K(X)$ can be written $E - n$, with $E \in \mathrm{Vect}(X)$ and n a positive integer thought of as a trivial bundle in $\mathrm{Vect}(X)$, and it enables one to extend 'clutching' from $\mathrm{Vect}(X)$ to $K(X)$.

Quillen associates to any category C with a composition-law like the direct sum in the category of vector spaces a functor K_C such that $K_C(X)$ is formed from the families of objects of C parametrized by X, and such that $K_C(X)$ is $K(X)$ if C is the category of finite dimensional complex vector spaces. For algebraic K-theory the main example is when C is the finitely generated projective modules over a discrete ring R, but the simplest example is the category of finite sets (with disjoint union as the composition-law), and I shall discuss it as it displays all the essential features of the general case.

A bundle of finite sets on a compact space X is to be interpreted as a covering space $Y \to X$ with finitely many sheets.

The isomorphism classes of these form a monoid (under disjoint union) $\mathrm{Cov}(X)$ which is a contravariant homotopy-functor of X and has the same clutching property as $\mathrm{Vect}(X)$. Its important difference from $\mathrm{Vect}(X)$ is that if one were to make it into an abelian group in the usual way--by considering formal differences $Y_1 - Y_2$ of finite coverings of X --the resulting functor would not have the Mayer-Vietoris property (b) above. (For a given covering Y there is usually no 'complementary' covering $Y^{\perp}$ such that $Y \amalg Y^{\perp}$ is trivial.) One deals with this by making $\mathrm{Cov}(X)$ into a group not considering each space X separately but treating the functor Cov as a whole. In general, writing $C(X)$ for the bundles of objects of a category C on X, one makes the

DEFINITION. The K-theory of a category C (with a composition-law) is a representable contravariant functor K_C from compact spaces to abelian groups together with a transformation of functors $C(X) \to K_C(X)$ preserving addition. It is characterized as universal among additive transformations from $C(X)$ into representable abelian-group-valued functors.

The K-groups of the category C are defined as the values of the functor K_C on spheres. Thus they are the homotopy groups of the space which represents K_C.

To justify the preceding definition one must see that the functor it characterizes exists. I shall outline the argument.

The first step is to observe that the functor $X \longmapsto C(X)$ is representable. This is a general fact--indeed the representing space is just the classifying-space [9] or realization BC of the category C --but in most cases it is clear directly. For example any real vector bundle $E \to X$ can be embedded as a sub-bundle in an infinite-dimensional trivial bundle $X \times \mathbb{R}^{\infty}$, and then assigning to $x \in X$ the fibre E_x at x gives a map from X to $\mathrm{Gr}(\mathbb{R}^{\infty})$, the Grassmannian of finite-dimensional subspaces of $\mathbb{R}^{\infty}$. This construction establishes an isomorphism $\mathrm{Vect}(X) \cong [X;\mathrm{Gr}(\mathbb{R}^\infty)]$. Similarly for a covering map $Y \to X$ one can choose $i : Y \to \mathbb{R}^\infty$ so that $p \times i : Y \to X \times \mathbb{R}^\infty$ is an embedding, and then taking $x \in X$ to the finite subset $i(p^{-1}(x))$ of $\mathbb{R}^\infty$ gives a map from X to the configuration space $C(\mathbb{R}^\infty)$ of finite subsets of $\mathbb{R}^\infty$. This leads to $\mathrm{Cov}(X) \cong [X;C(\mathbb{R}^\infty)]$. (In §3 I shall discuss the relation between $C(\mathbb{R}^\infty)$ and $\mathrm{Gr}(\mathbb{R}^\infty)$ and the categorical classifying spaces.)

The second step is to represent the 'stabilized' functor $X \mapsto C_{stab}(X)$, where $C_{stab}(X)$ consists of the formal differences $Y - n$, where $Y \in C(X)$ and n is an object of C thought of as a trivial bundle. I shall call the representing space for this $B_\infty C$. In the case of $\mathrm{Cov}(X)$ it can be constructed as the telescope of the sequence $C(\mathbb{R}^\infty) \to C(\mathbb{R}^\infty) \to C(\mathbb{R}^\infty) \to \ldots$, where each map adds a point to the configurations in a standard way, or alternatively, more abstractly, as $\mathbb{Z} \times B\Sigma_\infty$, where Σ_∞ is the infinite symmetric group.

With ordinary K-theory the process is now complete, because stable vector bundles form a group. But for covering spaces the space $B_\infty C$ has fundamental group Σ_∞, which is not abelian, so it cannot represent a group-valued functor. (It seems at first paradoxical that when X is compact $[X;B\Sigma_\infty]$ is naturally an abelian semigroup but usually not a group, although $B\Sigma_\infty$ is a connected space. If one considered base-point-preserving maps it would not be abelian.) Quillen has shown that in all cases one can construct a space $B^{ab}C$ representing an abelian-group-valued functor with a homology equivalence $B_\infty C \to B^{ab}C$. This will solve the universal problem of the definition in view of the theorem of obstruction theory that a homology equivalence $B \to B'$ induces a bijection $[B';Y] \to [B;Y]$ for any simple space Y. There are two ways of constructing $B^{ab}C$ in general. One is simply to attach 2- and 3-dimensional cells to $B_\infty C$ so as to make its fundamental group abelian without changing its homology. The other is to observe that BC is a topological monoid and so has a classifying space $B(BC)$; then $B^{ab}C$ can be defined as the loop-space $\Omega B(BC)$ of this. The second method gives a little more information, but requires the 'group-completion theorem' ([4],[7]), which asserts that $B_\infty C \to \Omega B(BC)$ is a homology isomorphism.

In some cases the functors K_C can be identified with better-known ones. For example when C is finite sets a theorem of Barratt, Priddy and Quillen asserts that K_C is stable cohomotopy, the universal cohomology theory, represented by $\varinjlim_k \Omega^k S^k$. (Cf. McDuff's talk.) When C is the vector spaces over a field with q elements K_C can be described as the functor which fits into a long exact sequence $\ldots \to K_C(X) \to K(X)[\frac{1}{q}] \xrightarrow{\psi^q - id} K(X)[\frac{1}{q}] \to \ldots$ where ψ^q is the Adams operation.

3. Classifying Spaces.

We have mentioned the identifications $Cov(X) \cong [X;C(\mathbb{R}^\infty)]$ and $Vect(X) \cong [X;Gr(\mathbb{R}^\infty)]$. But the systematic representing space for the bundles of objects of a category C on X is the classifying space BC of [9]. Recall that BC is constructed by taking a vertex for each object of C , adding a 1-simplex for each morphism, adding a 2-simplex for each commutative diagram

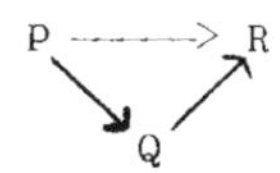

in C , and so on. Thus it is the realization of the simplicial object; $ob(C) \leftleftarrows mor(C) \leftleftarrows mor_2(C) \leftleftarrows \ldots$, where $mor_k(C)$ is the set of chains of k composable morphisms in C . There is little change needed if C is a topological category, i.e. if $ob(C)$ and $mor(C)$ (and hence all $mor_k(C)$) are topological spaces rather than sets. For BC is still defined as the space obtained by attaching all the spaces $\Delta^k \times mor(C)$ together in the obvious way. (Δ^k is the standard k-simplex.)

The categories corresponding to covering spaces and real vector bundles are finite sets and isomorphisms and finite dimensional real vector spaces and isomorphisms. The latter is a topological category: its objects form a discrete set, but for any two objects V, W the space of isomorphisms $V \to W$ has its usual topology.

The equivalence between $C(\mathbb{R}^\infty)$ and $Gr(\mathbb{R}^\infty)$ and the respective classifying-spaces follows simply from two principles. (Cf. [9] and [11] Appendix A.)

PROPOSITION (3.1). An equivalence of topological categories $F : C \to C'$, or more generally any functor F such that there exists a functor $G : C' \to C$ and transformations between $G \circ F$ and id_C and $F \circ G$ and $id_{C'}$ (in either direction), induces a homotopy equivalence $BC \to BC'$.

PROPOSITION (3.2). A functor $F : C \to C'$ which induces homotopy equivalences $ob(C) \to ob(C')$ and $mor(C) \to mor(C')$ induces a homotopy equivalence $BC \to BC'$.

Using these propositions one proves, for example, that $B(Vect) \simeq Gr(\mathbb{R}^\infty)$ as follows. Think of $Gr(\mathbb{R}^\infty)$ as a topological category with $Gr(\mathbb{R}^\infty)$ as its space of objects and no morphisms except identities. Then $B(Gr(\mathbb{R}^\infty)) = Gr(\mathbb{R}^\infty)$. Let $Vect/\mathbb{R}^\infty$ be the topological category whose objects are pairs (V, α) with $V \in ob(Vect)$ and $\alpha : V \to \mathbb{R}^\infty$ an embedding, and whose morphisms $(V, \alpha) \to (V', \alpha')$ are maps $\beta : V \to V'$ in Vect such that $\alpha = \alpha' \beta$. There are functors

$$\begin{array}{ccccc} Vect & \longleftarrow & Vect/\mathbb{R}^\infty & \longrightarrow & Gr(V) \\ V & \longleftarrow\!\shortmid & (V, \alpha) & \longmapsto & \alpha(V) . \end{array}$$

The right-hand functor is an equivalence of categories, so $B(Vect/\mathbb{R}^\infty) \xrightarrow{\simeq} B(Gr(\mathbb{R}^\infty)) = Gr(\mathbb{R}^\infty)$ by (3.1). The left-hand functor induces homotopy equivalences on objects and morphisms, for

$$\mathrm{ob}(\mathrm{Vect}/\mathbb{R}^\infty) = \coprod_V \mathrm{Emb}(V) \quad , \quad \mathrm{mor}(\mathrm{Vect}/\mathbb{R}^\infty) = \coprod_{V \to V'} \mathrm{Emb}(V') \ ,$$

where the topological sums are over the objects V and morphisms $V \to V'$ of Vect , and $\mathrm{Emb}(V)$ and $\mathrm{Emb}(V')$ denote the spaces of embeddings of V and V' in $\mathbb{R}^\infty$, which are contractible. So $B(\mathrm{Vect}/\mathbb{R}^\infty) \xrightarrow{\simeq} B(\mathrm{Vect})$ by (3.1), and so $B(\mathrm{Vect}) \simeq \mathrm{Gr}(\mathbb{R}^\infty)$. The case of finite sets and configuration spaces is exactly analogous.

The following two examples (due to Quillen) are more interesting. In §2 I explained how K-theory is obtained by passing from the topological semigroup BC to its 'group completion' $B^{ab}C$. But the group-completion process can be carried out on the category rather than the space, i.e. one can construct a category $\hat{C}$ such that $B\hat{C} \simeq B^{ab}C$. $\hat{C}$ is the category of 'virtual objects' of C . Its objects are pairs (V_0, V_1) of objects of C , and a morphism $(V_0, V_1) \to (V_0', V_1')$ is a triple $(W; \alpha_0, \alpha_1)$, where W is an object of C and $\alpha_i : V_i \oplus W \to V_i'$ are isomorphisms in C , but subject to the identification $(W; \alpha_0, \alpha_1) = (W'; \alpha_0', \alpha_1')$ if there is an isomorphism $\beta : W \to W'$ such that $\alpha_i'(\mathrm{id}_{V_i} \oplus \beta) = \alpha_i$.

Quillen also constructed a category QC such that the space of loops in B(QC) is equivalent to $B^{ab}C$. Its objects are those of C , but a morphism $V \to V'$ in QC is a triple $(W_+, W_-; \alpha)$, where W_+ and W_- are objects of C and $\alpha : W_+ \oplus V \oplus W_- \to V'$ is an isomorphism in C . Again $(W_+, W_-; \alpha) = (W_+', W_-'; \alpha')$ if there are $\beta_\pm : W_\pm \to W_\pm'$ such that $\alpha'(\beta_+ \oplus \mathrm{id}_V \oplus \beta_-) = \alpha$. (This is the version of QC appropriate for K-theory using direct sums rather than exact sequences; Quillen actually wrote about the latter in [8].)

When $C = \mathrm{Vect}$ then just as BC is related to the infinite Grassmannian the classifying spaces of $\hat{C}$ and QC correspond to the spaces of Fredholm operators and self-adjoint Fredholm operators in Hilbert space. The idea is that a Fredholm operator is determined by its kernel and cokernel--a pair of finite dimensional vector spaces--in the sense that the operators with a prescribed kernel and cokernel form a contractible space. When a Fredholm operator is deformed continuously its kernel and cokernel can jump, but only by adding isomorphic pieces to each: i.e. the jumps correspond to morphisms in $\hat{C}$. In the same sense a self-adjoint operator is determined by its kernel, and when the kernel jumps the piece added to it is the sum of a part on which the operator was positive and a part on which it was negative, so that the jump corresponds to a morphism in QC.

More formally let H be a real or complex Hilbert space, and let $\mathrm{Fred}(H)$ and $\mathrm{Saf}(H)$ be the space of Fredholm operators and self-adjoint Fredholm operators in H with the norm topology.

PROPOSITION (3.3). If $C = \mathrm{Vect}$ then

(a) $B\hat{C} \simeq \mathrm{Fred}(H)$,

(b) $B(QC) \simeq \mathrm{Saf}(H)$.

Proof of (a): First observe that $\hat{C}$ can be replaced by the category $\hat{C}'$, whose objects are morphisms $f : V_0 \to V_1$ of finite dimensional vector spaces and whose morphisms from $f : V_0 \to V_1$ to $f' : V_0' \to V_1'$ are pairs of embeddings $V_i \to V_i'$ such that the diagram

$$\begin{array}{ccc} V_0 & \xrightarrow{f} & V_1 \\ \downarrow & & \downarrow \\ V_0' & \xrightarrow{f'} & V_1' \end{array}$$

is commutative and cartesian, i.e. such that $\ker(f) \xrightarrow{\cong} \ker(f')$ and $\mathrm{coker}(f) \xrightarrow{\cong} \mathrm{coker}(f')$. This follows from (3.2) as the linear maps between vector spaces form a contractible space.

How, arguing as in the proof that $B(\mathrm{Vect}) \simeq \mathrm{Gr}(\mathbb{R}^\infty)$, and using as intermediate category the category of cartesian diagrams

$$\begin{array}{ccc} V_0 & \longrightarrow & V_1 \\ \downarrow & & \downarrow \\ H & \xrightarrow{f} & H \end{array}$$

where the vertical maps are embeddings and f is Fredholm, one sees that $\hat{BC}' \simeq BF$, where F is the ordered space of pairs (f,V), with $f \in \mathrm{Fred}(H)$ and V a finite dimensional subspace of H such that $V + \mathrm{im}(f) = H$, ordered by $(f,V) \leq (f',V')$ if $f = f'$ and $V \subset V'$.

The space BF maps to $\mathrm{Fred}(H)$, regarded as the space of a topological category with only identity morphisms, by $(f,V) \longmapsto f$. The inverse image of f under this map is the classifying space of the ordered space of all $V \subset H$ such that $V + \mathrm{im}(f) = H$, and is contractible because the ordered space is directed. To see that $BF \to \mathrm{Fred}(H)$ is actually a homotopy equivalence one needs a more delicate argument which does not seem worth including here. (A very similar argument is given in detail in (2.7) of [10].)

The proof of (3.3)(b) is similar. One considers vector spaces with an inner product, and begins by passing from QC to the category whose objects are pairs (V,f), where $f : V \to V$ is self-adjoint, and whose morphisms $(V,f) \to (V',f')$ are embeddings $\alpha : V \to V'$ such that $f'\alpha = \alpha f$ and $\alpha : \ker(f) \xrightarrow{\cong} \ker(f')$.

REFERENCES

1. J. F. Adams, A variant of E. H. Brown's representability theorem. Topology, 10 (1971), 185-198.

2. M. F. Atiyah, Bott periodicity and the index of elliptic operators. Quarterly J. Math. Oxford, 19 (1968), 113-140.

3. M. F. Atiyah and I. M. Singer, Index theory for skew-adjoint Fredholm operators. Publ. Math. Inst. des Hautes Etudes Scient. Paris, 37 (1969), 305-326.

4. M. G. Barratt and S. B. Priddy, On the homology of non-connected monoids and their associated groups. Comment. Math. Helvet., 47 (1972), 1-14.

5. E. H. Brown, Cohomology theories. Ann. of Math., 75 (1962), 467-484.

6. A. Dold and R. Thom, Quasifaserungen und unendliche symmetrische Produkte. Ann. of Math., 67 (1958), 239-281.

7. D. McDuff and G. Segal, Homology fibrations and the 'group completion' theorem. Inventiones math. 3. (1976), 279-284.

8. D. G. Quillen, Higher algebraic K-theory I. In: Algebraic K-theory I. Springer Lecture Notes 341 (1973).

9. G. B. Segal, Classifying spaces and spectral sequences. Publ. Math. Inst. des Hautes Etudes Scient. Paris, 34 (1968), 105-112.

10. G. B. Segal, Configuration spaces and iterated loop spaces. Inventiones math., 21 (1973), 213-221.

11. G. B. Segal, Categories and cohomology theories. Topology, 13 (1974), 293-312.

SOME REMARKS ON OPERATOR THEORY AND INDEX THEORY

I. M. Singer, Massachusetts Institute of Technology[1]

I. A TYPE II INDEX THEOREM.

§1. Statement and proof. Let D be an elliptic pseudo-differential operator, $D: C^\infty(E) \to C^\infty(F)$, E and F vector bundles over a compact smooth manifold M of dim n. The index theorem states that ind D (dim ker D - dim ker D^*) $= (-1)^n \int_{T(M)} \text{ch}\, \sigma_D\, \tau_M$. Here σ_D, the symbol of D, is interpreted as an element of $K(T(M))$ [2, p. 556], and $\text{ch}: K(T(M)) \to H^*(T(M), Q)$ is the Chern character; τ_M is the Todd class of $T(TM))$ over $T(M)$ which has an almost complex structure.

When V is another vector bundle over M, then one can define an elliptic operator $D\otimes I_V : C^\infty(E\otimes V) \to C^\infty(F\otimes V)$ so that $\sigma_{D\otimes I_V} = \sigma_D \otimes I_V$ and $\text{ch}\, \sigma_{D\otimes I_V} = \text{ch}\, \sigma_D \cdot \text{ch}\,(V)$. For example, let $\{f_u\}$ be a smooth partition of unity subordinate to a finite cover $\{U\} \ni V|_U \simeq U \times \mathbb{C}^N$. Let $D\otimes I_V = \sum_U D\otimes f_u I$ with $D\otimes f_u I$ well defined on compactly supported sections of $E\otimes V|_U$. For index theory the method of definition is not important. The point is there exists an operator with symbol $\sigma_D \otimes I_V$ and ind $D\otimes I_V = (-1)^n \int_{T(M)} \text{ch}\, \sigma_D \cdot \text{ch}\, V\, \tau_M$. If V is a flat bundle of dim N, then $\text{ch}\, V = N \cdot 1$ so that ind $D\otimes I_V = N \cdot$ ind D.

Our purpose is to extend the index theorem to vector bundles which are no longer finite dimensional in the usual sense; in fact they are Hilbert bundles which are finite dimensional in the Murray-von Neumann sense. To do this, we use the Breuer-Fredholm theory for rings of type II_∞, and for simplicity of exposition we will assume that $\mathcal{U}$ is a factor of type II_∞. Then $\mathcal{F}_{II}$ (the Fredholms of type II) $= \{T \in \mathcal{U} : \exists\, S \in \mathcal{U}$ with $TS-I, ST-I \in \mathcal{J}_{\mathcal{U}}\}$. Here

---

(1) Research supported by NSF Grant #MCS 75-23334 and #MPS 72-05055 A03.

$\mathcal{I}_{\mathcal{U}}$ is the unique nontrivial ideal of $\mathcal{U}$; it is the closure of the ideal generated by $[P \in \mathcal{U}\ ;\ P$ a projection $\dim_{II} P < \infty]$. Then one has $\mathrm{ind}_{II} : \mathcal{I}_{II} \xrightarrow{\text{cont}} \mathbb{R}$, $\mathbb{R}$ in the discrete topology where $\mathrm{ind}_{II}\ T = \dim_{II} \ker T - \dim_{II} \ker T^*$. See Breuer [4] for this and the next two paragraphs.

Hilbert bundles of finite $\dim_{II}$ can be constructed in various ways. For our purposes, we use the following. Let $\mathcal{M}$ be a II_1 factor such that $\mathcal{M} \hat{\otimes} I_\infty \simeq \mathcal{U}$ and suppose $\mathcal{M}$ is a factor operating on a Hilbert space H with $\mathcal{M}'$ also finite. Given a finite covering $\{U\}$ of M and transition functions $g_{UV} : U \cap V \to (\mathcal{M}')^r$, the invertible elements of $\mathcal{M}'$, one constructs a Hilbert bundle $\mathcal{H}$ in the usual way whose fibres are modeled by H as an $\mathcal{M}$-module.

In the standard case, it is well known that $\mathcal{I}$ is a classifying space for $K(X)$, i.e. $K(X) \simeq [X, \mathcal{I}]$ and the Bott periodicity theorem says that $K(S^2(X)) \simeq K(X)$. Here, too, one can define the Grothendieck group $K_{II}(M)$ of the semigroup of $\mathcal{M}$-finite Hilbert bundles and it turns out again that $K_{II}(X) \simeq [X, \mathcal{I}_{II}]$. The periodicity theorem holds and this implies $K_{II}(X) \simeq K(X) \otimes \mathbb{R}$.

Suppose now that $\mathcal{H}_j$ $j = 1,2$ are $\mathcal{M}$-finite Hilbert bundles over M . We want to define pseudodifferential operators $D : C^\infty(\mathcal{H}_1) \to C^\infty(\mathcal{H}_2)$, decide which are elliptic and state an index theorem. Locally, $\mathcal{H}_j|_U \simeq U \times H$. The Kohn-Nirenberg definition of pseudodifferential operators extends to our situation [10].

$u \to Pu(x) = \frac{1}{(2\pi)^n} \int_{R^n} e^{ix\cdot\xi} p(x,\xi) \hat{u}(\xi) d\xi$ where $p(x,\xi) \in \mathcal{M}'$, is C^∞ ,

and $\| \partial_x^\alpha \partial_\xi^\beta p(x,\xi) \| \le C_{\alpha,\beta,K} (1 + |\xi|)^{s-\beta}$ on compact sets $K \subset U$; u is a smooth compactly supported function with values in H . The usual functional calculus holds and the class of operators is invariant under coordinate change.

For more details concerning operator valued $p(x,\xi)$ see G. Luke [11]. There, however, $p(x,\xi)$ can be any bounded operator i.e. has values in a I_∞ so that operators of order-∞ need not be compact. Special assumptions about $p(x,\xi)$ must be made in that case. Here, operators of order-∞ are in the ideal $\mathcal{I}_{\mathcal{H}}$ as follows: $L_2(M,\mathbb{H})$ is an $\mathcal{M}$ module with commutator $\mathcal{H}$ (for $L_2(M,\mathbb{H}) \simeq L_2(M) \otimes H$, so that the commutator of $\mathcal{M}$ is $B(L_2(M)) \otimes \mathcal{M}' \simeq I_\infty \hat{\otimes} \mathcal{M}' \simeq \mathcal{H}$). An operator of order-∞, by Fourier transform is locally an integral operator with smooth kernel $k(x,y)$ with values in $\mathcal{M}'$, hence lies in $\mathcal{I} \hat{\otimes} \mathcal{M}'$ where $\mathcal{I}$ is the ideal of compact operators on $L_2(M)$. Since $\mathcal{I} \otimes \mathcal{M}' = \mathcal{I}_{\mathcal{H}}$, a smoothing operator in our theory lies in $\mathcal{I}_{\mathcal{H}}$.

An operator P of order s is elliptic if its principal symbol $\sigma_p = p_o(x,\xi)$ is an invertible element of $\mathcal{M}'$ for all $(x,\xi) \in T(M/C)$. The functional calculus guarantees and operator Q of order-s such that $PQ-I$ and $QP-I$ are smoothing, i.e. of order-∞. Hence an elliptic operator of order 0 lies in $\mathcal{J}_{II}$. As in the standard case, H^{-s} can be defined as the range of a positive elliptic operator of order s. An operator $P: C^\infty(M,\mathbb{H}_1) \to C^\infty(M,\mathbb{H}_2)$ of order s extends to a bounded operator: $H^s(\mathbb{H}_1) \to L_2(\mathbb{H}_2) = H^o(\mathbb{H}_2)$ and is an $\mathcal{M}$-module map. If P is elliptic, then P is a type II Fredholm operator from H^s to H^o. From the point of view of index theory, one need only deal with o^{th} order operators because P can be replaced $P|\Delta|^{-s}$ where $|\Delta|$ is a first order positive operator.

As in the classical case, the homotopy class of σ_p lies in $K_{II}(T(M)) \simeq K(T(M)) \otimes R$ and $ch\ \sigma_p \in H^*(T(M)\ R)$. We state the following results of Atiyah and the author.

<u>THEOREM</u>. <u>If</u> P <u>is elliptic</u>, <u>then</u> $\mathrm{ind}_{II} P = (-1)^n \int_{T(M)} ch\ \sigma_p \cdot \tau_M$.

Corollary. <u>Let</u> $D: C^\infty(E) \to C^\infty(F)$ <u>be an ordinary elliptic operator and let</u> $\mathbb{H}$ <u>be an</u> $\mathcal{M}$-finite <u>Hilbert bundle</u>. <u>Then</u>
$\mathrm{ind}_{II}\ D \otimes I_H = (-1)^n \int_{T(M)} ch\ \sigma_D \cdot ch\mathbb{H} \cdot \tau_M$.

We sketch the proof of the theorem. Of course, one could go back to one of the original proofs of the index theorem and carry along the additional structure. This would give a proof unifying the type I and II case. Instead, we assume the index theorem and use the fact that $K_{II}(X) \simeq K(X) \otimes \mathbb{R}$. It suffices then to check the theorem on operators of the type $D \otimes I_{\mathcal{H}}$ where D is an ordinary elliptic operator and $\mathcal{H}$ is the Hilbert bundle representing $r \in \mathbb{R}$, i.e., $\mathcal{H}$ is a trivial bundle $M \times H$ with the coupling constant between $\mathfrak{m}$ and $\mathfrak{m}'$ equal to r. But then $\dim_{II} \ker D \otimes I_{\mathcal{H}} = r \dim \ker D$ and similarly for D^*. Since $\mathrm{ch}\mathcal{H} = r$, the theorem is proved for $D \otimes I_{\mathcal{H}}$.

§2. An Example. Let $\widetilde{M}$ be the simply connected covering space of M, and suppose D is a differential elliptic operator so that it can be lifted to $\widetilde{D} : C^\infty(\widetilde{E}) \to C^\infty(\widetilde{F})$ where $\widetilde{E}$ and $\widetilde{F}$ are $\pi^*(E)$ and $\pi^*(F)$ respectively with $\pi : \widetilde{M} \to M$ and $\mathrm{vol}\,\widetilde{M} = \pi^* \mathrm{vol}\, M$. Note that $\widetilde{D}$ commutes with $\pi_1(M) = G$ acting as deck transformations on $\widetilde{M}$. Let ρ be the left regular representation of G on $\ell_2(G) = H$. This induces a Hilbert bundle $\mathcal{H}$ over M with $= \widetilde{M} \underset{G}{\times} H$ where G acts on $\widetilde{M}$ as deck transformations and on H via $\rho(G)$. Note that $\mathcal{H}$ is a module over the group algebra $\mathfrak{A}(G)$ via the right regular representation of G on H. When E and F are trivial one dimensional bundles, the Fubini theorem tells us that $L_2(\widetilde{M}) = L_2(M,\mathcal{H})$; i.e. a function $f \in L_2(\widetilde{M})$ is an L_2 section of $\mathcal{H}$ over M; for associated to f is the function $F : \widetilde{M} \to H$ defined by $F(\widetilde{m})(g) = f(\widetilde{m}g)$. Since $F(\widetilde{m}g_1)(g) = f(\widetilde{m}g_1 g) = F(\widetilde{m})(g_1 g)$, F transforms according to the left regular representation and gives a section s_f of $\mathcal{H}$. $f \in L_2(\widetilde{M})$ implies $s_f \in L_2(M,\mathcal{H})$, for $F(\widetilde{m}) \in \ell_2(G)$ a.e. Similarly for vector bundles E and F. We leave to the reader the verification that $\widetilde{D} : C^\infty(\widetilde{E}) \to C^\infty(\widetilde{F})$ is the same as $D \otimes I_{\mathcal{H}} : C^\infty(M, E \otimes \mathcal{H}) \to C^\infty(M, F \otimes \mathcal{H})$.

When G is finite, $\widetilde{M}$ is compact and it is easy to see that the $\operatorname{ind} \widetilde{D}/_{o(G)} = \operatorname{ind} D$ for $\mathcal{H}$ is a flat vector bundle of dimension $o(G)$. Thus $\operatorname{ind}_{\pi_1} D = \operatorname{ind} D$ where $\operatorname{ind}_{\pi_1}$ means index normalized by $o(G)$. Another way of putting this is that $\ker D$ and $\ker D^*$ are invariant under G, so that $\operatorname{ind} D$ lies in the character ring of G. The formula above says that this element is the regular representation of G occurring with multiplicity $\operatorname{ind} D$.

Whon G is infinite, the Breuer index theory via the Murray-von Neumann dimension function gives a way of counting the multiplicity of the regular representation as it occurs in $\ker \widetilde{D} - \ker \widetilde{D}^*$ in L_2. For simplicity, assume that all conjugacy classes of G but the trivial one are infinite. This happens when M has a metric with negative curvature [14]. In particular, when $M = {}_G\backslash{}^H/_K$, H a semisimple group K a maximal compact subgroup and G a uniformizing discrete subgroup. For example, M a Riemann surface of genus > 1, $H = \mathfrak{sl}(2, \mathbb{R})$. Then the von Neumann algebra $\mathfrak{M}$ generated by $L_1(G)$ acting on H is a factor (of type II_1) and $\mathcal{H}$ is an $\mathfrak{M}$-finite Hilbert bundle in the sense of Breuer discussed above. Since $\mathcal{H}$ is flat, $\operatorname{ch} \mathcal{H} = r \cdot 1$, $r \in R$. But since the coupling constant between $\mathfrak{M}$ and $\mathfrak{M}'$ is 1, $\operatorname{ch} \mathcal{H} = 1$. Since D is really $\widetilde{D} \otimes I_{\mathcal{H}}$, $\widetilde{D}$ is elliptic. The theorem of §2 implies the

COROLLARY. $\operatorname{ind}_{II} \widetilde{D} = \operatorname{ind} D$.

M. F. Atiyah [1] has given a direct proof with little use of rings of type II. In fact $\dim_{II} \ker D$ is expressed by Atiyah as follows: Let P be the projection of $L_2(\widetilde{E})$ on $\ker \widetilde{D}$. The Bergmann kernel theory tells us that the Schwartz distribution of P is given by a smooth function $p(\widetilde{x}, \widetilde{y})$, $\widetilde{x}, \widetilde{y} \in M$. Since $\widetilde{D}$ commutes with the action of G, $p(\widetilde{x}g, \widetilde{y}g) = p(\widetilde{x}, \widetilde{y})$ so that $p(\widetilde{x}, \widetilde{x})$ is a function $p(x)$ with values in $\operatorname{Hom}(E_x, E_x)$ on the compact space M. Then $\dim_{II} \ker \widetilde{D} \overset{\text{def}}{=} \int_M \operatorname{tr} p(x)$. It is not hard to show that in this case the $\dim_{II} P(L_2(M, E \otimes \mathcal{H}))$ is just $\int_M \operatorname{tr} p(x)$. The importance of the corollary is that $\operatorname{ind} D > 0 \Rightarrow \exists$ nontrivial L_2 solutions to $\widetilde{D}\widetilde{u} = 0$.

A simple example leads to an interesting problem. Suppose $D = d + \delta : C^{\infty}(\Lambda^{even}) \to C^{\infty}(\Lambda^{odd})$ so that $\operatorname{ind} D = \chi(M) = \operatorname{ind}_{II} \widetilde{D}$. Let $\widetilde{\beta}_p = \dim_{II} \ker \widetilde{D}^*\widetilde{D}$, $\widetilde{D}^*\widetilde{D}$ the Laplacian on L_2 p-forms on $\widetilde{M}$; i.e., $\widetilde{\beta}_p = \dim_{II}$ of the L_2 harmonic p-forms. Then $\chi(M) = \operatorname{ind}_{II} \widetilde{D} = \Sigma (-1)^p \widetilde{\beta}_p$ is an integer. Are there examples where $\widetilde{\beta}_p$ is not an integer? J. Dodziuk [6] has recently shown that $\widetilde{\beta}_p$ is an invariant of the covering $\pi : \widetilde{M} \to M$, by developing a simplicial L_2 theory and proving an L_2 - de Rham theory for $\widetilde{M}$.

§3. The Non Factor Case. When $\mathcal{H}$ is not a factor, then $\mathcal{H}$ is the direct integral $\int_X \mathcal{H}_x$, X the maximal ideal space of the center of $\mathcal{H}$ (or $\mathcal{M}$). Suppose $\mathcal{H}$ has no piece of type III; then $\mathcal{H} = \int_{X_I} \mathcal{H}_x + \int_{X_{II}} \mathcal{H}_x$ giving a splitting into type I and II. A Fredholm operator $F \in \mathcal{H}$ is $\int_X F_x$ and $\operatorname{ind} : F \to \operatorname{ind}_x F_x$ gives an element of $C(X_1, Z) \oplus C(X_2, \mathbb{R}^d)$, where $\mathbb{R}^d$ is the reals in the discrete topology. Since X is totally disconnected, the family of Fredholm operators $\{F_x\}$ is not interesting. However, in the flat bundle situation, i.e. the covering space example, $\pi : \widetilde{M} \to M$, the center of $\mathcal{M}$ is the weak closure of the center of the convolution algebra $\ell_1(\pi_1)$. It is generated by functions constant on finite conjugacy classes of π_1 . Moreover since the characteristic function χ_e of $e \in \pi_1$ lies in H , $\mathcal{M}$ has a natural trace: $T \to \langle T\chi_e, \chi_e \rangle$. Let C be the center of the C^* algebra generated by $\ell_1(\pi_1)$ and suppose its maximal ideal space is Y . Then $\widetilde{D}$ is a family of elliptic operators $\{D_y\}_{y \in Y}$. The natural trace induces a continuous linear functional on C and therefore a measure μ on Y . One can show that $\mathrm{ind}_{II}\tilde{D} = \int_Y \mathrm{ind}_y D_y \, \mu$ where ind_y is the index for the fibre over y. (Note that now ind_{II} is an abuse of notation for in fact each fibre may be of type I). When π_1 is abelian for example, $Y = \hat{\pi}_1$ and $\{D_y\}$ is a family of elliptic operators indexed by $\hat{\pi}$, the character group of π_1. In fact, $D_y = D \otimes I_{V_y}$, V_y the line bundle induced by y. The families index of $\{D_y\}$ turns out to have some applications [12]. When $\mathfrak{M}$ is purely of type II, $\{D_y\}$ is a family of type II Fredholm operators whose family index lies in $K_{II}(Y) = K(Y) \otimes \mathbb{R} \simeq H^{even}(Y, \mathbb{R})$. We know of no applications.

II. SOME GEOMETRIC PROBLEMS RELATED TO BDF THEORY.

§1. The Index Theorem. BDF have given an isomorphism of $\mathrm{EXT}(X)$ with $K_*(X)$, X a finite dimensional compact space. See Atiyah in this volume for an explicit map and [6,9] for $\mathrm{EXT}(X)$ as a homology theory. Dual to the Chern map $ch^* : K^*(X) \to H^*(X,Q)$, one has $ch_* : K_*(X) \to H_*(X,Q)$.

Douglas [7] exhibits a map $\phi : \mathrm{EXT}(X) \to \mathrm{Hom}(\tilde{K}^1(X), Z)$ whose kernel is the torsion subgroup, as follows: First, $\tilde{K}^1(X) = \lim[X, G\ell(N,\mathbb{C})]$. If $\alpha : o \to \mathcal{J} \to \mathfrak{a} \xrightarrow{\sigma} C(X) \to o$ represents an element of $\mathrm{EXT}(X)$, let $\alpha_N : o \to \mathcal{J} \otimes \mathfrak{M}_N \to \mathfrak{a} \otimes \mathfrak{M}_N \xrightarrow{\sigma_n} C(X) \otimes \mathfrak{M}_N \to o$ where $\mathfrak{M}_N = N \times N$ matrices $= \mathrm{Hom}(\mathbb{C}^N)$. Then $\mathcal{J} \otimes \mathfrak{M}_N \simeq$ the compact ideal on $H \otimes \mathbb{C}^N$. $\mathcal{J} \cap \mathfrak{a} \otimes \mathfrak{M}_N = \sigma^{-1} \, C(X, G\ell(N,\mathbb{C}))$. Since the index is continuous and depends only on the symbol, we get a map $\phi_N(\alpha) : [X, G\ell(N,\mathbb{C})] \xrightarrow{ind} Z$. This map is stable, defining ϕ. This map ϕ gives the natural pairing of $K_1(X) \otimes \tilde{K}'(X) \to Z$ corresponding to the pairing $H_*(X,Q) \otimes H^*(X,Q) \to Q$ under the Chern maps. That is, if $\sigma \in \tilde{K}^1(X)$, then $\phi(\alpha)(\sigma) = ch_*(\alpha)(ch^*\sigma)$. When α is given explicitly, an explicit expression for $ch_*(\alpha)$ can be viewed as an index theorem.

For example, suppose $\mathcal{P}$ is the algebra of oth order pseudo-differential operators on a compact smooth manifold M. One has the extension $\alpha: 0 \to \mathcal{J} \to \bar{\mathcal{P}} \to C(S(M)) \to 0$ where $S(M)$ is the tangent sphere bundle of M. If $T \in \bar{\mathcal{P}} \otimes \mathcal{M}_N$ is Fredholm, then the index theorem says $\text{ind}\, T = (-1)^n \int_{B(M)} ch(\delta(\sigma_T)) \cdot \tau_M$ [2, p. 602] where $H^*(S(M)) \overset{\delta}{\to} H^*(B(M), S(M))$. Stoke's theorem gives $\text{ind}\, T = (-1)^n \int_{S(M)} ch^*(\sigma_T)\, \tau_M|_{S(M)}$. Thus $ch_*(\alpha)$ is the homology dual of $\tau_M|_{S(M)}$.

Another example is obtained when X is the smooth boundary of a strictly pseudoconvex domain $\mathcal{D} \subset \mathbb{C}^N$. See Venugopalkrishna [16]. Let $H = L_2$-boundary values of functions holomorphic in $\mathcal{D}$ and let Q be the projection: $L_2(X) \to H$. Let $\mathfrak{A} = C^*$-algebra generated by the Toeplitz operators $[PM_f, M_f$ multiplication by $f \in C(X)]$. Then $\alpha: 0 \to \mathcal{J} \to \mathfrak{A} \to C(X) \to 0$ gives an element of $EXT(X)$. It turns out that $ch_*(\alpha) = 1$. This was shown by Venugopolkrishna when $\mathcal{D}$ is a ball and by Boutet de Monvel in general (private communication).

Note that in both cases, the space X is a contact manifold with natural line bundle a symplectic manifold. In the first case, the symplectic manifold is $T^*(M) \simeq T(M)$ and in the second case it is $L = T(X)/_{W(X)}$ where $W(X) = T(X) \cap i(T(X))$. Since the Todd class of $T(L)$ is 1 (every symplectic manifold is an almost complex manifold so the Todd class is defined), the two index theorems have the same form.

A unified treatment seems in order, especially since $ch_*(\alpha)$ is obtained in both cases from the finer structure of the space. In the first case, the symplectic structure is obtained from $\mathcal{P}^1$, the first order operators. $\mathcal{P}^1$ is a Lie algebra under Lie bracket which goes into Poisson bracket under the symbol map, and the Poisson bracket gives the symplectic structure of $T^*(\mathcal{M})$. In the second case, the symplectic structure comes from the complex structure of $\mathcal{D}$.

If (X, η) is a contact manifold, let $\pi: L \to X$ be the line bundle, $T(X)/_{\eta^{-1}(o)}$. Then L has a symplectic structure given by the two forms $d\pi^*(\eta)$. Ideas of Howe suggest that associated to (X,η) is an element $\alpha \in EXT(X)$ or $EXT(\widetilde{X})$, $\widetilde{X}$ a double cover, and one conjectures that $ch_*(\alpha)$ is the homology dual of the Todd class of $T(L)$ restricted to $S(L)$. The two examples given should be extreme cases in the sense that the first has a totally real polarization and the second a complex polarization in $T(L)$.

§2. PL Manifolds. If X is a PL manifold of dim 4k , one can define the (unstable) L-polynomial essentially in terms of the signature of submanifolds [13]. Automatically, sign $X = \int_X L(X)$. But more generally, for any $\xi \in K^o(X)$, the map $\xi \to \int_X L(X)\, ch\, \xi$ is a homomorphism of $K^o(X) \to Z$. In this volume, Atiyah exhibits a surjective map $\beta : E\ell\ell(X) \to K_o(X)$ and one has the natural surjective map of $K_o(X) \to Hom(K^o(X),Z)$. Hence there exists an elliptic operator $D \in E\ell\ell(X)$ such that ind $D\otimes I_\xi = \int_X L(X)\, ch\, \xi$. Can one find a concrete realization of such a D ? When X is smooth, $D = {}^{d\delta-\delta d}/{d\delta+\delta d+1} : C^\infty(\Lambda^+) \to C^\infty(\Lambda^-)$. When X is PL , $\Lambda^\pm$ do not exist as vector bundles, but $L_2(\Lambda^\pm)$ are definable. Is D definable directly as a singular integral operator? See [15] for a variation of this question.

§3. Singular Varieties. Baum, Fulton, and McPherson [3,8] have proved a Riemann-Roch theorem for varieties X . Let $K_o^{alg}(X)$ denote the Grothendieck group generated by coherent analytic sheaves over X . They exhibit a homology class τ_X and a map $\Phi_X : K_o^{alg}(X) \to K_o(X)$, such that if E is a holomorphic vector bundle, then the Euler characteristic of the cohomology of X with coefficients in the sheaf of germs of E is given by $(\tau_X \cap ch(E))_o$ where $(\ )_o$ is the oth component of the homology class; and if θ is the structure sheaf of X , $\tau_X = ch_*\Phi_X(\theta)$ is the "Todd class" of

X . Since $\Phi_X(\theta) \in K_o(X)$, there exists a $D \in E\ell\ell(X) \ni \Phi_X(\theta) = \beta(D)$. When X is nonsingular, then $D = \bar{\partial}^* + \bar{\partial} / \Delta^{\frac{1}{2}} + I : C^\infty(\Lambda^{o,\text{even}} \otimes E) \to C^\infty(\Lambda^{o,\text{odd}} \otimes E)$ by the Index theorem. In the singular case, is there a geometric realization of D ? More generally if S is a coherent analytic sheaf, there exists a $D_1 \ni \beta(D_1) = \Phi_X(S)$. Is there a geometric realization of the elliptic operator D_1 ?

§4. Invariance of the Rational Pontrjagin Classes. Let C_j , $j = 1,2$ be two smooth structures on the compact manifold X and let $\mathcal{P}_J$ denote the corresponding algebras of 0th order pseudo-differential operators, giving extensions $\alpha_j \in \text{EXT}(X) : o \to \mathcal{J} \to \bar{\mathcal{P}}_j \to C(S(X)) \to o$. (The sphere bundles can be identified.) The index theorem together with the Novikov theorem on the invariance of the rational Pontrjagin classes imply that $\alpha_1 - \alpha_2$ is a torsion element in $\text{EXT}(X)$, i.e. $\phi(\alpha_1) = \phi(\alpha_2)$ in $\text{Hom}(\tilde{K}_1(X), Z)$. One of the original motivations of the use of operator algebras in index theory was the hope that $\bar{\mathcal{P}}_1$ was connected to $\bar{\mathcal{P}}_2$ by a path of extensions, thereby proving the Novikov theorem. The existence of torsion in $\text{EXT}(X)$ shows this may not be possible. But for some N , $N\bar{\mathcal{P}}_1$ is pathwise connected to $N\bar{\mathcal{P}}_2$. Can this be shown directly?

REFERENCES

1. M. F. Atiyah, Elliptic operators, discrete groups, and von Neumann algebras, to appear.

2. M. F. Atiyah and I. M. Singer, The index of elliptic operators III, Ann. of Math., 87 (1968), 546-604.

3. P. Baum, W. Fulton, and R. MacPherson, Riemann-Roch for singular varieties, Publ. Math. I.H.E.S. no. 45 (1975).

4. M. Breuer, Fredholm theories in von Neumann algebras I and II, Math. Ann., 178 (1968), 243-254; and 180 (1969), 313-325.

5. _________, Theory of Fredholm operators and vector bundles, Rocky Mountain Math. Consortium 3 (1973), 383-429.

6. J. Dodziuk, deRham theory for L^2 cohomology of infinite coverings, to appear.

7. R. G. Douglas, The relation of EXT to K-theory, to appear.

8. W. Fulton, A Hirzebruch-Riemann-Roch formula for analytic spaces and non-projective algebraic varieties, to appear.

9. J. Kaminker and C. Schochet, K-theory and Steenrod homology; applications to the BDF theory of operator algebras, to appear in TAMS.

10. J. J. Kohn and L. Nirenberg, An algebra of pseudo-differential operators, Comm. Pure Appl. Math., 18 (1965), 269-305.

11. G. Luke, Pseudodifferential operators on Hilbert bundles, Journal of Diff. Equations, 12 (1972), 566-589.

12. G. Lusztig, Novikov's higher signature and families of elliptic operators, J. Diff. Geom., 7 (1972), 229-256.

13. J. Milnor and J. Stashev, Characteristic classes, Princeton Univ. Press and Univ. of Tokyo Press (1974).

14. A. Preissman, Quelques proprieties globales des espaces de Riemann, Comment. Math. Helv., 15 (1942-3), 175-216.

15. I. M. Singer, Future extensions of index theory and elliptic operators, Prospects in Mathematics, Princeton University Press and Univ. of Tokyo Press (1973), 171-185.

16. U. Venugopalkrishna, Fredholm operators associated with strongly pseudoconvex domains in $\mathbb{C}^n$, J. Functional Anal., 9 (1972), 349-372.

FACTORS OF TYPE III

Masamichi Takesaki, University of California, Los Angeles

Today, the structure of a factor of type III is described as follows:

THEOREM 1. *Every factor $\mathcal{M}$ of type III is isomorphic to the crossed product $W^*(\mathcal{N},\mathbb{R},\theta)$ of a uniquely associated covariant system $\{\mathcal{N},\theta\}$ of a von Neumann algebra $\mathcal{N}$ of type II_∞ and a one parameter automorphism group $\{\theta_t : t \in \mathbb{R}\}$ such that the restriction of θ to the center $\mathcal{C}$ of $\mathcal{N}$ is ergodic, but not isomorphic to the translation on $L^\infty(\mathbb{R})$, and θ transforms some faithful semi-finite normal trace τ on $\mathcal{N}$ in such a way that $\tau \circ \theta_t = e^{-t}\tau$, $t \in \mathbb{R}$. Here the uniqueness of $\{\mathcal{N},\theta\}$ means that if $\{\mathcal{N}_1,\theta^1\}$ and $\{\mathcal{N}_2,\theta^2\}$ are covariant systems satisfying the conditions for $\{\mathcal{N},\theta\}$, then $W^*(\mathcal{N}_1,\mathbb{R},\theta^1) \cong W^*(\mathcal{N}_2,\mathbb{R},\theta^2)$ is equivalent to the conjugacy of $\{\mathcal{N}_1,\theta^1\}$ and $\{\mathcal{N}_2,\theta^2\}$ in the sense that there exists an isomorphism π of $\mathcal{N}_1$ onto $\mathcal{N}_2$ such that $\theta_t^2 = \pi \circ \theta_t^1 \circ \pi^{-1}$, $t \in \mathbb{R}$.* cf. [2], [8], [12], [13], [28] and [29].

The aim of this paper is to present the background of the above result together with some of further development. Although it is impossible to elaborate here, I would like to emphasize that the recent interaction between mathematics and theoretical physics was indispensable in this achievement.

In 1967, there were two very important achievements in the theory of operator algebras: R. Powers distinguished a continuum of non-isomorphic factors of type III [23] and M. Tomita showed that given a von Neumann algebra $\mathcal{M}$ on a Hilbert space $\mathcal{H}$ with separating cyclic vector ξ_0 then there exist a conjugate linear unitary involution J and a non-singular positive self-adjoint operator Δ such that

(i) $J\Delta^{\frac{1}{2}}x\xi_0 = x^*\xi_0$, $x \in \mathcal{M}$;

(ii) $J\mathcal{M}J = \mathcal{M}'$ and $\Delta^{it}\mathcal{M}\Delta^{-it} = \mathcal{M}$, $t \in \mathbb{R}$.

After Power's work, a rapid progress in the classification theory of

* This is supported in part by N. S. F.

factors followed: Araki and Woods classified the factors obtained as infinite tensor product of finite factors of type I, abbreviated as ITPFI factor, by introducing algebraic invariants $r_\infty(\mathcal{M})$ and $\rho(\mathcal{M})$ in 1968 [3] and McDuff constructed continua of factors of type II_1 and II_∞ in 1969, [20], which was also confirmed by Sakai, [24]. The developments along this line was treated in a new book, by Anastasio and Willig. [1]

A quiet but steady development followed after Tomita's work, [30]. A serious inspecting seminar on Tomita's work took place and confirmed his result at the University of Pennsylvania for 1968/69, which was later published as lecture notes [26] by the present author. The major discovery in the seminar was that the one parameter automorphism group $\{\sigma_t\}$ of $\mathcal{M}$ given by $\sigma_t(x) = \Delta^{it} x \Delta^{-it}$, $t \in \mathbb{R}$ and $x \in \mathcal{M}$, and the normal functional φ given by $\varphi(x) = (x\xi_0 \mid \xi_0)$, $x \in \mathcal{M}$, satisfy the Kubo-Martin-Schwinger condition: for any pair $x,y \in \mathcal{M}$ there exists a continuous bounded function $F(z)$ on $0 \leq \mathrm{Im} z \leq 1$, holomorphic inside the strip such that

$F(t) = \varphi(\sigma_t(x)y)$ and $F(t+i) = \varphi(y\sigma_t(x))$, and that $\{\sigma_t\}$ is uniquely determined by φ subject to the KMS condition; hence it is denoted by $\{\sigma_t^\varphi\}$ and called the modular automorphism group. The notion of the KMS-condition came from physics as the name suggests. Haag, Hugenholtz and Winnink showed in 1967, [16], that the cyclic representation π_φ of a C^*-algebra A induced by a state φ satisfying the KMS-condition with respect to a given one parameter automorphism group $\{\sigma_t\}$ on A is standard: there exists a unitary involution J such that $J\,\pi_\varphi(A)''\,J = \pi_\varphi(A)'$ and $JaJ = a^*$, $a \in \pi(A)'' \cap \pi(A)'$. It is widely believed that an equilibrium state in quantum statistical mechanics is characterized by the KMS-condition.

As an illustration, let us consider an example. A faithful normal positive linear functional φ on the algebra $\mathcal{L}(\mathcal{H})$ of all bounded operators is given by

$$\varphi(x) = \mathrm{Tr}(xh),\ x \in \mathcal{L}(\mathcal{H}),$$

with some non-singular positive operator h of the trace class. If $\dim \mathcal{H} < +\infty$ and $h = \lambda 1$, $\lambda > 0$, then we have $\varphi(xy) = \varphi(yx)$ for every $x, y \in \mathcal{L}(\mathcal{H})$, that is, φ is a trace. If this is the case, then the involution $x \to x^*$ in $\mathcal{L}(\mathcal{H})$ is a unitary involution J in the Hilbert space structure in $\mathcal{L}(\mathcal{H})$ induced by φ, which gives rise to a symmetry between the left multiplication representation and the right multiplication representation of $\mathcal{L}(\mathcal{H})$ on this Hilbert space $\mathcal{L}(\mathcal{H})$. In general, $\varphi(xy) \neq \varphi(yx)$ because $x h \neq hx$. However, xh and hx are homotopic under the homotopy: $t \in [0,1] \to h^t x h^{1-t}$. An analytic expression of this homotopy is nothing but the KMS-condition, that is, if we consider the one parameter automorphism group $\sigma_t(x) = h^{it} a h^{-it}$, then the $\mathcal{L}(\mathcal{H})$-valued function $f(t) = h\sigma_t(x)$ is extended analytically to the strip, $0 \leq \mathrm{Im} z \leq 1$; and we have

$$f(t) = h\sigma_t(x) \text{ and } f(t+i) = \sigma_t(x)h.$$

Thus, we see that the KMS-condition or the modular automorphism group measures and compensates the non-trace like behavior of φ. As a matter of fact, we have the following characterization:

THEOREM 2. A σ-finite von Neumann algebra $\mathcal{M}$ is semi-finite if and only if the modular automorphism group $\{\sigma_t^\varphi\}$ of a faithful normal positive linear functional φ on $\mathcal{M}$ is implemented by a one parameter unitary group $\{u(t)\}$ in $\mathcal{M}$. If the predual $\mathcal{M}_*$ is separable, then the innerness of each individual automorphism σ_t^φ is sufficient for $\mathcal{M}$ to be semi-finite. (cf. [22] and [26]).

This result mildly indicates some connection between the algebraic structure of the von Neumann algebra $\mathcal{M}$ in question and the behavior of the modular automorphism group.

There was another fortunate mature development in the theory of operator algebras. In 1966, G.K. Pedersen proposed a simultaneous generalization of positive linear functionals and semi-finite traces on a C^*-algebra under the terminology C^*-integrals, which was further generalized by F. Combes to the notion of weights on a C^*-algebra. (cf. [5] and [21]). It turns out that the combination of the theory of weights and the KMS-condition is very sueful in the study of the structure of von Neumann algebras.

DEFINITION 3. A <u>weight</u> on a von Neumann algebra $\mathcal{M}$ is a map φ of the positive cone $\mathcal{M}_+$ to the extended positive reals $[0,\infty]$ such that

$$\varphi(x + y) = \varphi(x) + \varphi(y),\ x,\ y \in \mathcal{M}_+;$$

$$\varphi(\lambda x) = \lambda\varphi(x),\ \lambda \geq 0,$$

with the usual convention $0(+\infty) = 0$. A weight φ is said to be <u>normal</u> if $\varphi(\sup x_i) = \sup \varphi(x_i)$ for every bounded increasing net $\{x_i\}$ in $\mathcal{M}_+$; <u>semi-finite</u> if $\mathfrak{n}_\varphi = \{x : \varphi(x^*x) < +\infty\}$ is σ-weakly dense in $\mathcal{M}$; <u>faithful</u> if $\varphi(x) > 0$ for every non-zero $x \in \mathcal{M}_+$.

A weight here means, however, always a faithful semi-finite normal one. Through Tomita's theory of modular Hilbert algebras, F. Combes showed, [6], that any weight φ on $\mathcal{M}$ gives rise to a unique one parameter automorphism group $\{\sigma_t^\varphi\}$ for which φ satisfies the KMS-condition in the sense that for any pair $x,y \in \mathfrak{n}_\varphi \cap \mathfrak{n}_\varphi^*$ there exists a continuous bounded function F on the strip, $0 \leq \operatorname{Im} z \leq 1$, holomorphic inside such that $F(t) = \varphi(\sigma_t^\varphi(x)y)$ and $F(t+i) = \varphi(y\sigma_t^\varphi(x))$ and that $\varphi \circ \sigma_t^\varphi = \varphi$, where one should note that φ is extended to a linear functional, denoted by φ again, on the linear span $\mathfrak{m}_\varphi$ of $\{x \in \mathcal{M}_+ : \varphi(x) < +\infty\}$ which agrees with $\mathfrak{n}_\varphi^*\mathfrak{n}_\varphi = \{y^*x : x,y \in \mathfrak{n}_\varphi\}$. Then Theorem 2 holds for weights without the restriction of σ-finiteness.

Investigating the relation between the Araki-Woods classification of ITPFI factors and the KMS-conditions, A. Connes showed in 1971 that asymptotic ratio set $r_\infty(\mathcal{M})$ of an ITPFI factor $\mathcal{M}$ is indeed the intersection of the spectrum $S_p(\Delta_\varphi)$ of the all possible modular operators Δ_φ; thus introduced a new algebraic invariant, the modular spectrum:

$$S(\mathcal{M}) = \cap\{S_p(\Delta_\varphi) : \varphi \text{ runs all weights on } \mathcal{M}\}.$$

He and Van Daele then showed in 1972 that $S(\mathcal{M})\setminus\{0\}$ is a closed subgroup of the multiplications group $\mathbb{R}_+^*$ if $\mathcal{M}$ is a factor; thus a new classification of factors of type III. A factor $\mathcal{M}$ is said to be of type III_λ, $0 < \lambda < 1$, if $S(\mathcal{M}) = \{\lambda^n : n \in \mathbb{Z}\} \cup \{0\}$; of type III_0 if $S(\mathcal{M}) = \{0,1\}$; of type III_1 if $S(\mathcal{M}) = \mathbb{R}_+^*$. Therefore, the factors distinguished by R. Powers were indeed factors of type III_λ, $0 < \lambda < 1$, with $\lambda = \frac{\mu}{1-\mu}$, where μ, $0 < \mu < \frac{1}{2}$, is a number defining a state ω_μ on the 2 x 2 matrix algebras by

$$\omega_\mu\left(\begin{bmatrix} x_{11}, & x_{12} \\ x_{21}, & x_{22} \end{bmatrix}\right) = \mu x_{11} + (1 - \mu)x_{22}.$$

In 1971, A. Connes further proved that the Araki-Woods invariant $\rho(\mathcal{M})$ for an ITPFI factor $\mathcal{M}$ is given under a trivial change of scale by the modular period group:

$$T(\mathcal{M}) = \{t \in \mathbb{R} : \sigma_t^\varphi = \iota \text{ for some weight } \varphi\},$$

and that $T(\mathcal{M})$ is a subgroup of the additive group $\mathbb{R}$. The formula between $\rho(\mathcal{M})$ and $T(\mathcal{M})$ for an ITPFI factor $\mathcal{M}$ is given by

$$\rho(\mathcal{M}) = \{e^{t/2\pi} : t \in T(\mathcal{M})\}.$$

By definition, $T(\mathcal{M})$ is an algebraic invariant for a factor $\mathcal{M}$. If $\mathcal{M}_*$ is separable, the semi-finiteness of $\mathcal{M}$ is equivalent to $T(\mathcal{M}) = \mathbb{R}$.

Besides these algebraic invariants, he showed the following:

THEOREM 4. [8] _If φ and ψ are weights on a von Neumann algebra $\mathcal{M}$, then there exists a unique σ-weakly continuous one parameter family $\{u_s\}$ of unitaries in $\mathcal{M}$ such that_

$$u_{s+t} = u_s\sigma_s^\varphi(u_t);$$

$$\sigma_t^\psi = \mathrm{Ad}\, u_t \circ \sigma_t^\varphi, \quad t \in \mathbb{R};$$

for any $x \in n_\varphi^* \cap n_\psi$ and $y \in n_\psi^* \cap n_\varphi$ there is a bounded continuous function F on the strip, $0 \leq \mathrm{Im}\, z \leq 1$, holomorphic inside the strip such that

$$F(t) = \varphi(\sigma_t^{\varphi}(x)u_t y), \quad F(t+i) = \psi(y\sigma_t^{\varphi}(x)u_t).$$

The construction of $\{u_t\}$ is surprisingly simple. Consider the weight χ on the tensor product $\mathcal{P} = \mathcal{M} \otimes \mathcal{B}_2$ on $\mathcal{M}$ and the 2 x 2 matrix algebra $\mathcal{B}_2$ defined by:

$$\chi\left(\begin{bmatrix} x_{11}, & x_{12} \\ x_{21}, & x_{22} \end{bmatrix}\right) = \varphi(x_{11}) + \psi(x_{22}).$$

It is then shown that

$$\sigma_t^{\chi}\left(\begin{bmatrix} 0, & 0 \\ 1, & 0 \end{bmatrix}\right) = \begin{matrix} 0, & 0 \\ u_t, & 0 \end{matrix}, \quad t \in \mathbb{R}.$$

If $\mathcal{M}$ is abelian, then φ and ψ are given by measures μ and ν on a Borel space Ω, and mutually absolutely continuous with respect to each other. Let $h = \frac{d\nu}{d\mu}$ be the Radon derivative of ν with respect to μ. Then $\{u_t\}$ is nothing but $\{h^{it}\}$. With this evidence, $\{u_t\}$ is called the <u>cocycle Radon-Nikodym derivative</u> of ψ with respect to φ and denoted by

$$u_t = (D\psi : D\varphi)_t, \quad t \in \mathbb{R}.$$

Considering the 3 x 3-matrix algebra over $\mathcal{M}$, he showed the chain rule:

$$(D\psi : D\varphi)_t = (D\psi : D\omega)_t (D\omega : D\varphi)_t, \quad t \in \mathbb{R},$$

for any three weights φ, ω and ψ.

It is clear from Connes' Radon-Nikodym theorem that

$$T(\mathcal{M}) = \{t \in \mathbb{R} : \sigma_t^{\varphi} \in \mathrm{Int}(\mathcal{M})\},$$

where $\mathrm{Int}(\mathcal{M})$ denotes the group of inner automorphisms; hence $T(\mathcal{M})$ is a subgroup of $\mathbb{R}$. He then showed that for any fixed weight φ on a factor $\mathcal{M}$,

$$S(\mathcal{M}) = \cap\, S_p(\Delta_{\varphi_e}),$$

where e runs over the central projections of the fixed point subalgebra $\mathcal{M}_{\varphi}$ of $\mathcal{M}$ under σ^{φ} and φ_e means the restriction of φ to $e\mathcal{M}e$. We call $\mathcal{M}_\varphi$ the <u>centralizer</u> of φ.

In order to get some idea about the structure of a factor of type III, let us consider a very special case. Suppose that a factor $\mathcal{M}$ admits a faithful normal state φ such that $\mathcal{M}_\varphi$ is a factor and $\sigma_T^\varphi = \iota$ for some $T > 0$. The smallest such $T > 0$ is called the <u>period</u> of φ. Connes proved, however, that every factor of type III_λ, $0 < \lambda < 1$, with separable predual admits such a state with $T = -2\pi/\log \lambda$. [8]). Let $\lambda = e^{-2\pi/T}$ and

$$\mathcal{M}_n = \{x \in \mathcal{M} : \sigma_t^\varphi(x) = \lambda^{int} x\}.$$

Of course, $\mathcal{M}_0 = \mathcal{M}_\varphi$. Clearly, we have

$$\mathcal{M}_n \mathcal{M}_m \subset \mathcal{M}_{n+m}, \ \mathcal{M}_n^* = \mathcal{M}_{-n}, \ n, m \in \mathbb{Z}.$$

Hence each $\mathcal{M}_n$ is a two-sided module over $\mathcal{M}_0$. It is not hard to see that $\mathcal{M}_1 \neq \{0\}$. Let $a = uh$ be the polar decomposition of an $a \in \mathcal{M}_1$. Then we have $u^*u = e \in \mathcal{M}_0$ and $uu^* = f \in \mathcal{M}_0$ and

$$\varphi(uxu^*) = \lambda\varphi(xu^*u), \ x \in \mathcal{M},$$

by the KMS-condition. Consider the tensor product $\bar{\mathcal{M}} = \mathcal{M} \otimes \mathcal{B}$, $\bar{\varphi} = \varphi \otimes \mathrm{Tr}$, $\bar{u} = u \otimes 1$, $\bar{e} = e \otimes 1$ and $\bar{f} = f \otimes 1$ where $\mathcal{B}$ denotes a factor of type I_∞. The centralizer $\bar{\mathcal{M}}$ of $\bar{\varphi}$ is $\mathcal{M}_\varphi \otimes \mathcal{B}$; hence infinite. The projections $\bar{e}$ and $\bar{f}$ are both infinite in $\mathcal{M}_{\bar{\varphi}}$, so that there are partial isometries v and w in $\mathcal{M}_{\bar{\varphi}}$ such that $v^*v = \bar{e}$, $vv^* = 1$, $ww^* = \bar{f}$ and $w^*w = 1$. Put

$$U = v\bar{u}w.$$

It follows that U is a unitary in $\bar{\mathcal{M}}$ such that $\sigma_t^{\bar{\varphi}}(U) = \lambda^{it} U$ and $U \bar{\mathcal{M}}_0 U^* = \bar{\mathcal{M}}_0$. Thus Ad U gives rise to an automorphism θ of $\bar{\mathcal{M}}_0$ and

$$\bar{\varphi} \circ \theta(x) = \lambda\bar{\varphi}(x), \ x \in \bar{\mathcal{M}}_0.$$

It is then easily shown that $\bar{\mathcal{M}} \cong W^*(\bar{\mathcal{M}}_0 \mathbb{Z}, \theta)$. Hence in this case, $\mathcal{M}$ is isomorphic to the crossed product of a factor $\bar{\mathcal{M}}_0$ by a single automorphism θ multiplying the trace by λ. The existence of such an automorphism implies that $\bar{\mathcal{M}}_0$ must be of type II_∞. Apart from the uniqueness, this is, in essence, the decomposition theorem for some factors of type III, at some earlier stage of the development in the structures theory in 1972. (cf. [2], [8] and [28]). The uniqueness requires similar Fourier analysis of the cocycle Radon-Nikodym derivatives $(D\psi : D\varphi)$. Instead of doing this, we will, however, go on to the general case.

Let $\mathcal{B}$ denote the algebra $\mathcal{L}^2(\mathbb{R})$ of square integrable functions on the real line $\mathbb{R}$ with respect to the Lebesgue measure. We define then one parameter unitary groups $\{U(t)\}$ and $\{V(s)\}$ in $\mathcal{B}$ by the following:

$$U(t)\ \xi(s) = \xi(s + t);$$
$$V(t)\ \xi(s) = e^{ist}\ \xi(s),\ \ \xi \in L^2(\mathbb{R}),\ s,\ t \in \mathbb{R}.$$

It follows then that

$$U(s)\ V(t)\ U(s)^*\ V(t)^* = e^{ist}\ 1,\ s,\ t \in \mathbb{R}.$$

Hence the one parameter automorphism groups $\{\mathrm{Ad}\ U(s)\}$ and $\{\mathrm{Ad}\ V(t)\}$ of $\mathcal{B}$ commute. Now, let $\mathcal{M}$ be a properly infinite von Neumann algebra. It is easily seen almost by definition that $\mathcal{M} \cong \mathcal{M} \otimes \mathcal{B}$. For a weight φ on $\mathcal{M}$, we consider the one parameter automorphism groups $\{\sigma_t\}$ and $\{\theta_t\}$ of $\mathcal{M} \otimes \mathcal{B}$ given by:

$$\begin{cases} \sigma_t = \sigma_t^{\varphi} \otimes \mathrm{Ad}\ U(t),\ t \in \mathbb{R}; \\ \theta_t = \iota \otimes \mathrm{Ad}\ V(t). \end{cases}$$

Clearly $\{\sigma_t\}$ and $\{\theta_s\}$ commute, so that $\{\theta_s\}$ gives rise to a one parameter automorphism group, denoted by $\{\theta_s\}$ again, of the fixed point algebra $\mathcal{N}$ of $\{\sigma_t\}$. It is not hard to see that $\mathcal{N}$ is generated by $1 \otimes U(t)$ and the operators:

$$\pi^{\varphi}(x)\ \xi(t) = \sigma_{-t}^{\varphi}(x)\ \xi(t),\ t \in \mathbb{R}, x \in \mathcal{M},\ \xi \in L^2(\mathcal{H};\mathbb{R}),$$

where $\mathcal{H}$ denotes a Hilbert space on which $\mathcal{M}$ acts; hence $\mathcal{N} \cong W^*(\mathcal{M},\mathbb{R},\sigma^{\varphi})$. We have the following:

LEMMA 5. <u>The von Neumann algebra $\mathcal{N}$ admits a faithful semi-finite normal trace τ such that $\tau\circ\theta = e^{-s}\tau$, $s \in \mathbb{R}$. The von Neumann algebra</u>

$\mathcal{M} \otimes \mathcal{B}$, *hence* $\mathcal{M}$, *is isomorphic to the crossed product* $W^*(\mathcal{N}, \mathbb{R}, \theta)$ *of* $\mathcal{N}$ *by* $\mathbb{R}$ *with respect to the action* θ *of* $\mathbb{R}$.

THEOREM 6. [29]. *If* $\mathcal{N}$ *is a von Neumann algebra equipped with a one parameter automorphism group* $\{\theta_s\}$ *and a faithful semi-finite normal trace* τ *such that* $\tau \circ \theta_s = e^{-s}\tau$, *then* (i) *the crossed product* $\mathcal{M} \quad W^*(\mathcal{N}, \mathbb{R}, \theta)$ *is properly infinite;* (ii) *the center* $\mathcal{C}_{\mathcal{M}}$ *of* $\mathcal{M}$ *is precisely the fixed point subalgebra* $\mathcal{C}_{\mathcal{N}}^{\theta}$ *of the center* $\mathcal{C}_{\mathcal{N}}$ *of* $\mathcal{N}$; *hence* $\mathcal{M}$ *is a factor if and only if* θ *is ergodic on the center* $\mathcal{C}_{\mathcal{N}}$ *of* $\mathcal{N}$; (iii) $\mathcal{M}$ *is semi-finite if and only if* $\mathcal{C}_{\mathcal{N}}$ *contains a continuous one parameter unitary group* $\{v(t)\}$ *such that* $\theta_s(v(t)) = e^{ist}\, v(t)$, $s, t \in \mathbb{R}$; (iv) *if* $\mathcal{M}$ *is of type* III, *then* $\mathcal{N}$ *is of type* II_∞ *and* $\{\mathcal{C}_{\mathcal{N}}, \theta\}$ *has no direct summand isomorphic to a multiple of* $L^\infty(\mathbb{R})$ *equipped with the translation;* (v) *if* $\{\mathcal{C}_{\mathcal{M}}, \theta\}$ *is ergodic, then*

$$S(\mathcal{M}) = \{e^t : \theta_t|_{\mathcal{C}_{\mathcal{N}}} = \iota\};$$

$$T(\mathcal{M}) = \{t \in \mathbb{R};\ \text{there exists a unitary } v \in \mathcal{C}_{\mathcal{N}}$$

with

$$\theta_s(v) = e^{ist}v,\ s \in \mathbb{R}\}.$$

As a direct consequence of Theorem 4, we have the following:

LEMMA 7. *Suppose that* $\{\mathcal{N}_1, \theta^1\}$ *and* $\{\mathcal{N}_2, \theta^2\}$ *are properly infinite von Neumann algebras equipped with one parameter automorphism groups and faithful semi-finite traces* τ_1 *and* τ_2 *respectively such that* $\tau_1 \circ \theta_s^1 = e^{-s}\tau_1$ *and* $\tau_2 \circ \theta_s^2 = e^{-s}\tau_2$, $s \in \mathbb{R}$. *Then the following two statements are equivalent:*

(i) $W^*(\mathcal{N}_1, \mathbb{R}, \theta^1) \cong W^*(\mathcal{N}_2, \mathbb{R}, \theta^2)$;

(ii) *There exist an isomorphism* π *of* $\mathcal{N}_1$ *onto* $\mathcal{N}_2$ *and a continuous one parameter family* $\{v_s\}$ *in* $\mathcal{N}_1$ *such that*

$$v_{s+t} = v_s\, \theta_s^1(v_t);\ s, t \in \mathbb{R};$$

$$\theta_s^2 = \pi \cdot \mathrm{Ad}(v_s) \circ \theta_s^1 \circ \pi^{-1}.$$

However, the next result, together with the above lemma, yields the uniqueness criteria in Theorem 1:

THEOREM 8. [13]. (i) *If θ is an automorphism of $\mathcal{N}$ such that $\tau\circ\theta\leq\lambda\tau$ for some $\lambda > 0$ and a faithful semi-finite normal trace τ. Then every unitary $u \in \mathcal{N}$ is of the form $v^*\theta(v)$ for some unitary $v \in \mathcal{N}$;*

(ii) *If $\{\theta_s\}$ is a one parameter automorphism group of $\mathcal{N}$ such that $\tau \circ \theta_s = \lambda^s\tau$ for some $\lambda \neq 1$ and a faithful semi-finite normal trace τ, then for every θ-one cocycle $\{u_s\}$, that is, a continuous one parameter family of unitaries in $\mathcal{N}$ with $u_{s+t} = u_s\ \theta_s(u_t)$, there exists a unitary $v \in \mathcal{N}$ such that $u_s = v^*\theta_s(v)$, $s \in \mathbb{R}$.*

Here a natural question is how this structure theorem and the discrete decomposition described above are related. The answer is quite simple: If $\mathcal{M}$ is of type III_λ, $0 < \lambda < 1$, then $\{\mathcal{C}_{\mathcal{N}},\theta\}$ is periodic with the period $T = -2\pi/\log\lambda$. Hence, considering the central decomposition

$$\mathcal{N} = \int_\Gamma^\oplus \mathcal{N}(\gamma)d\mu(\gamma),$$

θ_T induces an antomorphism of each fibre algebra $\mathcal{N}(\gamma)$. The covariant systems $\{\mathcal{N}(\gamma),\theta_T\}$ are equivalent to the appearing in the above discrete decomposition. In the type III_0 case, A. Connes proved the following:

THEOREM 9. [8]. *If $\mathcal{N}$ is a von Neumann algebra with non-atomic center and equipped with an automorphism θ and a faithful semi-finite normal trace τ such that $\tau \circ \theta \leq \lambda\tau$ for some $0 < \lambda < 1$ and θ is ergodic on the center $\mathcal{C}_{\mathcal{N}}$ of $\mathcal{N}$, then the crossed product $\mathcal{M} = W^*(\mathcal{N},\theta)$ of $\mathcal{N}$ by θ is a factor of type III_0. Every factor of type III_0 is of this form for some $\{\mathcal{N},\theta\}$.*

The uniqueness criteria for factors of type III_0 requires more preparations; so we omit the detail. But he did give the uniqueness of this decomposition within some equivalence.

Once again examining the way the II_∞-von Neumann algebra $\mathcal{N}$ was constructed, one realizes that the algebra $\mathcal{N}$ is the centralizer of the weight $\bar{\omega} = \varphi \otimes \omega$ on $\mathcal{M} \otimes \mathcal{B}$ where the weight ω on $\mathcal{B}$ is given by

$$\omega(x) = \mathrm{Tr}(hx),\ x \in \mathcal{B}_+;$$
$$h = \exp(\frac{d}{dt}),$$

i.e. $(D\omega : D\mathrm{Tr})_t = U(t),\ t \in \mathbb{R}$. A. Connes proved indeed, in the course of proving the converse of the cocycle Radon-Nikodym theorem, that for any one-cocycle $\{u_t\}$ in $\mathcal{M}$, there exists a unitary $v \in \mathcal{M} \otimes \mathcal{B}$ such that

$$u_t \otimes U(t) = v^* \sigma_t(v),\ t \in \mathbb{R}.$$

In other words, for any weights φ and ψ on $\mathcal{M}$, $\psi \otimes \omega$ and $\varphi \otimes \omega$ are conjugate under the inner automorphism group $\mathrm{Int}(\mathcal{M}\otimes\mathcal{B})$. This means then that on a properly infinite von Neumann algebra there is a unique class of weights which describes the structure of the algebra. The weights of this class is characterized by the following:

THEOREM 10. [13]. _Let $\mathcal{M}$ be an infinite factor with separable predual. For a weight $\bar{\omega}$ on $\mathcal{M}$ with properly infinite centralizer, the following two conditions are equivalent:_

(i) _For any $\lambda > 0$, there exists a unitary $u \in \mathcal{M}$ such that_ $\lambda\bar{\omega}(x) = \bar{\omega}(uxu^*),\ x \in \mathcal{M}_+;$

(ii) _For any weight φ on $\mathcal{M}$, there exists an isomorphism π of $\mathcal{M}$ onto $\mathcal{M} \otimes \mathcal{B}$ such that_

$$\bar{\omega}(x) = (\varphi\otimes\omega) \circ \pi(x),\ x \in \mathcal{M}_+,$$

where ω is the weight on $\mathcal{B}$ defined above.

DEFINITION 11. [13]. The weight $\bar{\omega}$ satisfying the condition in the above theorem is called _dominant_.

In other words, a dominant weight is characterized as one fixed, within unitary equivalence, under the multiplication by positive scalars.

Let $\mathcal{W}_{\mathcal{M}}$ denote the space of all weights on $\mathcal{M}$ and $\mathcal{W}_{\mathcal{M}}^0$ the space of all weights with properly infinite centralizer. For a pair φ, ψ of weights on $\mathcal{M}$, we write $\varphi \prec \psi$ if there exists an isometry $u \in \mathcal{M}$ with $uu^* \in \mathcal{M}_\psi$ such that $\varphi(x) = \psi(uxu^*),\ x \in \mathcal{M}_+$. If the above u is unitary, then we write $\varphi \sim \psi$. We see then that "$\sim$" is equivalence relation associated with the partial ordering "$\prec$". The space $\mathcal{W}^0_{\mathcal{M}}/$"$\sim$" is then a σ-complete Boolean lattice which is isomorphic to the lattice of all σ-finite projections of a unique abelian von Neumann algebra $\underset{\sim}{\mathcal{P}}(\mathcal{M})$. To each $\varphi \in \mathcal{W}^0_{\mathcal{M}}$, there corresponds a unique projection $p(\varphi)$ of $\underset{\sim}{\mathcal{P}}(\mathcal{M})$ such that

$$\varphi \prec \psi \Longleftrightarrow p(\varphi) \leq p(\psi).$$

Since the multiplication by a positive scalar preserves the ordering, to each $\lambda > 0$ there corresponds a unique automorphism $\mathcal{T}^{\mathcal{M}}_\lambda$ of $\underset{\sim}{\mathcal{P}}(\mathcal{M})$ such that

$$\mathcal{T}^{\mathcal{M}}_\lambda p(\varphi) = p(\lambda\varphi), \quad \varphi \in \mathcal{W}^0_{\mathcal{M}}.$$

We call $\{\underset{\sim}{\mathcal{P}}(\mathcal{M}), p, \mathcal{T}_\lambda\}$ the <u>global flow of weights</u>. Theorem 10 means that there exists the only one σ-finite projection $d \in \underset{\sim}{\mathcal{P}}(\mathcal{M})$ invariant under $\mathcal{T}^{\mathcal{M}}_\lambda$, which is given by $d = p(\bar{\omega})$. Putting $p(\varphi) = p(\varphi \otimes \mathrm{Tr})$ for the general $\varphi \in \mathcal{W}_{\mathcal{M}}$, we have the following:

THEOREM 12. [13]. <u>Let $\mathcal{M}$ be an infinite factor with separable predual. For any $\varphi \in \mathcal{W}_{\mathcal{M}}$, the following conditions are equivalent</u>:

i) $\varphi \prec \bar{\omega}$;

ii) <u>The map</u>: $\lambda \in \mathbb{R}^*_t \to \mathcal{T}_\lambda p(\varphi) \in \underset{\sim}{\mathcal{P}}(M)$ <u>is σ-strongly continuous</u>;

iii) <u>The integral</u> $\int_{-\infty}^{\infty} \sigma^\varphi_t(x)dt = E_\varphi(x)$, $x \in \mathcal{M}_+$, <u>exists for σ-weakly dense</u> x's <u>in</u> $\mathcal{M}_+$.

DEFINITION 13. A weight φ is said to be <u>integrable</u> if φ satisfies any of the above conditions.

Therefore, $(\underset{\sim}{\mathcal{P}}_{\mathcal{M}})_d$ is the continuous part of the flow $\mathcal{T}^{\mathcal{M}}_\lambda$. The restriction of $\{\mathcal{T}^{\mathcal{M}}_\lambda\}$ to $(\underset{\sim}{\mathcal{P}}_{\mathcal{M}})_d = \mathcal{P}_{\mathcal{M}}$ is called the <u>smooth flow of weights</u> on $\mathcal{M}$, and denoted by $\{F^{\mathcal{M}}_\lambda\}$. Since there is no non-trivial invariant projection properly majorized by d, the smooth flow of weights is ergodic. By construction, the association: $\mathcal{M} \mapsto F^{\mathcal{M}}$ of the smooth flow of weights to each infinite factor $\mathcal{M}$ is a functor. The relation between this function $F^{\mathcal{M}}$ and the structure theorem, Theorem 6, is described as follows:

THEOREM 14. [13]. Let $\mathcal{M}$ be an infinite factor with separable predual and $\{\mathcal{N}, \theta\}$ be the covariant system over $\mathbb{R}$ in Theorem 6 such that $\mathcal{M} \cong W^*(\mathcal{N}, \mathbb{R}, \theta)$.

(i) $\{\mathcal{C}_{\mathcal{N}}, \theta_{-\mathrm{Log}\lambda}\} \cong \{\mathcal{O}(\mathcal{M}), F_\lambda^{\mathcal{M}}\}$;

(ii) $S(\mathcal{M})\setminus\{0\} = \{\lambda \in \mathbb{R}_+^* : F_\lambda^{\mathcal{M}} = \iota\}$.

Therefore, the algebraic invariant $S(\mathcal{M})$, the modular spectrum, of $\mathcal{M}$ is essentially the kernel of the smooth flow $F^{\mathcal{M}}$ of weights. One should note here that the smooth flow $F^{\mathcal{M}}$ of weights is defined directly, hence functionally, from $\mathcal{M}$. We then determine this flow for a factor given by the so-called group measure space construction.

Let $\mathcal{A}$ be an abelian von Neumann algebra with separable predual equipped with a continuous action α of a separable locally compact group G. This is equivalent to having a standard measure space $\{\Gamma, \mu\}$ equipped with a Borel action of G, and $\mathcal{A} = L^\infty(\Gamma, \mu)$, $\alpha_g(a)(\gamma) = a(g^{-1}\gamma)$, $a \in \mathcal{A}$, $g \in G$, $\gamma \in \Gamma$. For simplicity, we assume that the action of G is free in the sense that $N_g = \{\gamma : g\gamma = \gamma\}$ is a null set for every $g \neq e$, although this restriction is not necessary, cf [17]. Let $\mathcal{M} = W^*(\mathcal{A}, G, \alpha)$. If the action of G is ergodic, then $\mathcal{M}$ is a factor. We have then the following:

(i) $\mathcal{M}$ is of type I $\Longleftrightarrow$ The action of G on Γ is transitive;

(ii) $\mathcal{M}$ is of type II_1 $\Longleftrightarrow$ The action of G on Γ is not transitive and admits a finite invariant measure;

(iii) $\mathcal{M}$ is of type II_∞ $\Longleftrightarrow$ The action is not transitive and admits an infinite invariant measure;

(iv) $\mathcal{M}$ is of type III $\Longleftrightarrow$ The action does not admit any invariant measure,

where the measures here are absolutely continuous with respect to the original measure μ. Let ρ be a positive Borel function on $G \times \Gamma$ such that

$$\int f(g\gamma)\rho(g,\gamma)d\mu(\gamma) = \int f(\gamma)d\mu(\gamma);$$

$$\rho(g_1g_2,\gamma) = \rho(g_1,g_2\gamma)\rho(g_2,\gamma),$$

namely $\rho(g,\gamma) = \frac{d\mu\circ g}{d\mu}(\gamma)$. Consider the product measure space $\Gamma \times \mathbb{R}_+^*$, where $\mathbb{R}_+^*$ is equipped with the Lebesgue measure m. By setting

$$T_g(\gamma,\lambda) = (g\gamma, \rho(g,\gamma)\lambda),\ \gamma \in \Gamma,\ \lambda > 0;$$

$$\Phi_\mu(\gamma,\lambda) = (\gamma,\lambda\mu),$$

G and $\mathbb{R}_+^*$ act on $\Gamma \times \mathbb{R}_+^*$ and commute. Hence we get an abelian von Neumann algebra $L^\infty(\Gamma\times\mathbb{R}_+^*, \mu\times m)$ on which G and $\mathbb{R}_+^*$ act.

THEOREM 15. [13]. _In the above situation, the smooth flow $F^{\mathcal{M}}$ of weights on $\mathcal{M}$ is isomorphic to the action of $\mathbb{R}_+^*$ on the fixed point subalgebra $L^\infty(\Gamma\times\mathbb{R}_+^*)^G$ induced naturally by $\{\Phi_\lambda\}$._

This construction is known as the _Anzai skew product_, or the closure of the range of the module ρ by G.W. Mackey [18]. A recent result of W. Krieger, [17], can be interpreted in the following way:

THEOREM 16. [17]. _In the same situation as above, if G is the additive integer group $\mathbb{Z}$, or equivalently if the action is given by a single ergodic transformation, then the smooth flow $F^{\mathcal{M}}$ of weights on $\mathcal{M}$ is a complete invariant for the algebraic structure of $\mathcal{M}$._

Thus, we have the following equivalence in different problems:
"The weak equivalence classification of the ergodic transformations"
∼"The classification of the factors given by the group measure space construction from an ergodic transformation"
∼"The conjugacy classification of the ergodic flows".

The weak equivalence classification of ergodic transformation groups was first introduced by H.A. Dye [15], and he proved in fact that all countable abelian ergodic transformation groups with finite invariant measure are weakly equivalent and give rise to hyperfinite II_1-factors. This classification was later reformulated by G.W. Mackey as the isomorphism classification of virtual subgroups. The relation between the weak equivalence classification of ergodic transformation groups and the isomorphism classification of the associated factors has been puzzled since Dye's work. In fact, H. Choda showed that if an isomorphism of the two factors associated with ergodic transformation groups preserves the maximal abelian subalgebras canonically attached to the constructions, then the groups are indeed weakly equivalent. [4].

The conjugacy classification of ergodic transformations and flows is, of course, one of the central problems in ergodic theory. Apparently, the weak equivalence classification looks much coarser than the conjugacy classification. But the above mentioned fact says that they are indeed the same problem.

Unlike the discrete crossed product, the relative commutant of the original algebra in the crossed product behaves mysteriously in general. We do have, however, the following:

THEOREM 17. [13]. _If φ is an integrable weight on a factor $\mathcal{M}$ with separable predual, then the relative commutant $\mathcal{M}_\varphi' \cap \mathcal{M}$ of the centralizer $\mathcal{M}_\varphi$ of φ is contained in $\mathcal{M}_\varphi$ as the center $\mathcal{C}_\varphi$._

This result, together with the construction of antomorphisms similar to that of [25], enables us to prove the following:

THEOREM 18. [13]. _Let φ be an integrable weight on a factor $\mathcal{M}$ with separable predual. There exists an isomorphism $\bar{\sigma}^\varphi$ of the multiplicative group $Z^1(F^{\mathcal{M}})$ of unitary one co-cycles of the smooth flow of weights on $\mathcal{M}$ onto the group of all automorphisms leaving the centralizer $\mathcal{M}_\varphi$ elementwise fixed, such that $\bar{\sigma}^\varphi_{\bar{t}} = \sigma^\varphi_t$, where $\bar{t} \in Z^1(F^{\mathcal{M}})$ means the cocycle given $\bar{t}(\lambda) = \lambda^{it}$, $\lambda > 0$, and $\bar{\sigma}^\varphi_c$, $c \in Z^1(F^{\mathcal{M}})$, is inner if and only if c is cobundary, i.e. there exists a unitary $v \in \mathcal{C}(\mathcal{M})$ such that $c_\lambda = v^* F^{\mathcal{M}}_\lambda(v)$, $\lambda > 0$._

Therefore, this extended modular automorphism group $\{\bar{\sigma}^\varphi_c : c \in Z^1(F^{\mathcal{M}})\}$ can be viewed as the Galois group of $\mathcal{M}$ relative to $\mathcal{M}_\varphi$. Furthermore, the co-cycle Radon-Nikodym derivative $\{(D\psi:D\varphi)_t : t \in \mathbb{R}\}$ is extended to $\{(D\psi : D\varphi)_c : c \in Z^1(F^{\mathfrak{M}})\}$, which behaves in the obvious way with respect to $\{\bar{\sigma}_c^{\varphi}\}$ and $\{\bar{\sigma}_c^{\psi}\}$. Hence there exists an isomorphism $\bar{\delta}_{\mathfrak{M}}$ independent of φ, of $H^1(F^{\mathfrak{M}}) = Z^1(F^{\mathfrak{M}})/B^1(F^{\mathfrak{M}})$ into $\mathrm{Out}(\mathfrak{M}) = \mathrm{Aut}(\mathfrak{M})/\mathrm{Int}(\mathfrak{M})$. Fixing the decomposition $\mathfrak{M} = W^*(\mathfrak{N}, \mathbb{R}, \theta)$ in Theorem 6, we can obtain an exact sequence:

$$\{1\} \to H^1(F^{\mathfrak{M}}) \xrightarrow{\bar{\delta}_{\mathfrak{M}}} \mathrm{Out}(\mathfrak{M}) \xrightarrow{\bar{\gamma}} \mathrm{Out}_{\tau,\theta}(\mathfrak{N}) \to \{1\},$$

where $\mathrm{Out}_{\tau,\theta}(\mathfrak{N}) = \{\varepsilon(\alpha) : \alpha \in \mathrm{Aut}(\mathfrak{N}),\ \tau \circ \alpha = \tau,\ \alpha\theta_s = \theta_s\alpha\}$ and ε means the canonical homomorphism of $\mathrm{Aut}(\mathfrak{N})$ onto $\mathrm{Out}(\mathfrak{N})/\mathrm{Int}(\mathfrak{N})$.

We should note here that the extended modular automorphism $\bar{\sigma}_c^{\varphi}$ is, in some sense, "functional calculus" of the "generator" of the modular automorphism group $\{\sigma_t^{\varphi}\}$. The evidence of this fact is the following: If $\mathfrak{M}$ is a semi-finite factor then $F^{\mathfrak{M}}$ is isomorphic to $L^{\infty}(\mathbb{R}^*)$ with translations; hence every $c \in Z^1(F^{\mathfrak{M}})$ is of the form $c_\lambda = f\, F_\lambda^{\mathfrak{M}}(f^*)$, $f \in L^{\infty}(\mathbb{R}^*)$ and if $\varphi = \mathrm{Tr}(h\cdot)$, then $\bar{\sigma}_c^{\varphi} = \mathrm{Ad}(f(h))$.

The smooth flow $F^{\mathfrak{M}}$ of weights being a functor, each $\alpha \in \mathrm{Aut}(\mathfrak{M})$ gives rise to an automorphism $\mathrm{mod}(\alpha)$ of the flow $F^{\mathfrak{M}}$ by

$$\mathrm{mod}(\alpha)p(\varphi) = p(\varphi \circ \alpha^{-1}),$$

which corresponds in the semi-finite case to the translation by λ determined by $\tau \circ \alpha = \lambda\tau$. Thus we call mod the <u>fundamental</u> <u>homomorphism</u> after Murray and von Neumann. We leave the detail to the original paper [13].

After all, the problem in understanding the structure of von Neumann algebras is reduced to the von Neumann algebras of type II_1 and type II_∞. Here, A. Connes has been making some substantial progress especially in the analysis of automorphism groups. cf [9] and [10]. The author believes that we will be able to understand much better the structure of von Neumann algebras in the near future.

REFERENCES

1. S. Anastasio and P.M. Willig, The Structure of Factors, Algonthnic Press, New York, 1974.

2. H. Araki, Structure of some von Neumann algebras with isolated discrete modular spectrum.

3. H. Araki and E.J. Woods, A classification of factors, Publications of Research Institute for Math. Sciences, Kyoto Univ., Ser. A 4 (1968), 51-130.

4. H. Choda, On the crossed product of abelian von Neumann algebras, II. Proc. Japan Acad. 43 (1967), 198-201.

5. F. Combes, Poids sur une C^*-algebra, J. Math. pures et oppl. 47 (1968), 57-100.

6. ________, Poids associe une algebre hilbertinne a gauche, Compotio Math. 23 (1971), 49-77.

7. A. Connes, Un nouvel invariant pour les algebres de von Neumann, C.R. Acad. Paris, Ser. A 273 (1971), 900-903; Calcul des deux invariants d'Araki et Woods par la theorie de Tomita et Takesaki, C.R. Acad. Paris, Ser. A 274 (1972), 175-177.

8. ________, Une classification de facteurs de type III, Ann. Sci. Ecole Norm. Sup. 4 eme Ser. 6 (1973), 133-252.

9. ________, Periodic automorphisms of the hyperfinite factor of type II_1, preprint.

10. ________, Automorphism groups of II_1-factors, Talk given at the International Conference on C^*-algebras and Applications to physics held in Rome, March, 1975.

11. ________ and A. Van Daele, The group property of the invariant S(M), Math. Scand.

12. ________ and M. Takesaki, Flots des poids sur les facteurs de type III, C. R. Acad., Paris, Ser. A 278 (1974), 945-948.

13. A. Connes and M. Takesaki, The flow of weights on factors of type III, to appear.

14. A. Van Daele, A new approach to the Tomita-Takesaki theory of generalized Hilbert algebras, J. Functional Analysis 15 (1974), 378-393.

15. H.A. Dye, On groups of measure preserving transformations, I, American J. Math. 81 (1959), 119-159; II, American J. Math. 85 (1963), 551-576.

16. R. Haag, N.M. Hugenholtz and M. Winnik, On the equilibrium states in quantum statistical mechanics, Comm. Math. Phys. 5 (1967), 215-236.

17. W. Krieger, An ergodic flows and the isomorphism of factors, to appear.

18. G.W. Mackey, Ergodic theory and virtual groups, Math. Ann. 166 (1966), 187-207.

19. D. McDuff, A countable infinity of II_1-factors, Ann. Math. 90 (1969), 361-371.

20. ________, Uncountable many II_1-factors, Ann. Math. 90 (1969), 372-377.

21. G.K. Pedersen, Measure theory for C^*-algebras, Math. Scand. 19 (1966), 131-145.

22. G.K. Pedersen and M. Takesaki, The Radon-Nikodym theorem for von Neumann algebras, Acta Math. 130 (1973), 53-87.

23. R.T.Powers, Representations of uniformly hyperfinite algebras and their associated von Neumann rings, Ann. Math. 86 (1967), 138-171.

24. S. Sakai, An uncountable number of II_1 and II_∞ factors, J. Functional Analysis 5 (1970), 236-246.

25. I.M. Singer, Automorphisms of finite factors, American J. Math. 77 (1955), 117-133.

26. M. Takesaki, Tomita's theory of modular Hilbert algebras, Lecture Notes in Math., Springer-Verlag 128 (1970).

27. __________, Periodic and homogeneous states on a von Neumann algebras, I, Bull. Amer. Math. Soc. 79 (1973), 202-206; II, Bull. Amer. Math. Soc. 79 (1973), 416-420; III, Bull. Amer. Math. Soc. 79 (1973), 559-563.

28. __________, The structure of a von Neumann algebra with a homogeneous periodic state, Acta Math. 131 (1973), 79-121.

29. __________, Duality for crossed products and the structure of von Neumann algebras of type III, Acta Math. 131 (1973), 249-310.

30. M. Tomita, Standard forms of von Neumann algebras, the Vth Functional Analysis Symposium of the Math. Soc. of Japan, Sendai, (1967).

TWISTED PRODUCTS OF BANACH ALGEBRAS AND THIRD ČECH COHOMOLOGY[1]

Joseph L. Taylor, University of Utah

Let A be a commutative Banach algebra with identity and let Δ be its maximal ideal space. Suppose one has somehow come across some information about the topology of Δ. What does this information say about the structure of A ? This is the question at the heart of a line of inquiry in Banach algebra theory that began with Shilov twenty three years ago and is still being actively pursued.

The prototype theorems in this subject are the Shilov Idempotent theorem [14] and the Aren-Royden Theorem [1,13]. The former characterizes the Čech group $H^0(\Delta,Z)$ as the additive subgroup of A generated by idempotents, while the latter characterizes $H^1(\Delta,Z)$ as $A^{-1}/\exp(A)$. In each case, the theorem is trivial if A is $C(\Delta)$, but non-trivial for general algebras in the sense that the proof rests on the Shilov-Arens-Calderon holomorphic functional calculus.

Arens [2] generalized the Arens-Royden Theorem by proving that the group consisting of $GL_n(A)$ modulo its identity component is isomorphic to $[\Delta, GL_n(\mathbb{C})]$ -- the group of homotopy classes of maps from Δ to $GL_n(\mathbb{C})$. Eidlin [4] and Novodvorskii [10] pointed out that this result can be used to characterize the group $K^1(\Delta)$ of Atiyah-Hirzebruch K-theory in terms of the structure of A. In the same paper, Novodvorskii proved that $K^0(\Delta)$ is isomorphic to $K_0(A)$, where $K_0(A)$ is the Grothendieck group of algebraic K-theory for A. Forster [5] proved that the second Čech group $H^2(\Delta,Z)$ is isomorphic to the Picard group $\mathrm{pic}(A)$.

It turns out that one can formulate a general theorem which has each of the above results as a simple consequence: If M is a locally closed complex submanifold of $\mathbb{C}^n$, then M determines a set $A_M \subset A^n$. If A is semi-simple then A_M is just

[1] Research sponsored by the National Science Foundation under NSF Grant No. GP-43114X.

$\{\alpha = (a_1,\ldots,a_n) \in A^n : \sigma(\alpha) \subset M\}$, where $\sigma(\alpha)$ is the joint spectrum of α . If A is not semi-simple then A_M is defined in a more complicated way (cf. [16]). In any case, if M has the structure of a complex homogeneous space (a complex Lie group modulo a closed complex Lie subgroup), then the Gelfand transform induces a bijection $[A_M] \to [\Delta, M]$, where $[A_M]$ is the set of components of A_M and $[\Delta, M]$ is the set of homotopy classes of maps from Δ to M . As stated here, this theorem is proved in [16], however the idea originates with Novodvorskii [10] where he proves the result in the case where M is open in $\mathbb{C}^n$.

The above result may be used whenever one wishes to relate to A invariants of Δ given by a functor F on compact spaces which has the form $F(\Delta) = [\Delta, M]$ for a space M which is approximable (in the homotopy sense) by complex homogeneous spaces. The space M is called a classifying space for F . The functors $H^0(\cdot,Z)$, $H^1(\cdot,Z)$, $H^2(\cdot,Z)$, $K^0(\ )$, and $K^1(\cdot)$ all have the right sort of classifying spaces and so the theorem applies. The same thing is true of the functors of real and symplectic K-theory (cf. [16]).

Now in attempting to characterize the higher (degree > 2) Čech cohomology groups of Δ , we ran into the following problem: apparently for $p > 2$, there is no classifying space for $H^0(\cdot,Z)$ approximable by complex homogeneous spaces. However, we also discovered that $H^3(\cdot,Z)$ has a classifying space which is a Banach complex homogeneous space (a complex Banach Lie group modulo a closed complex Banach Lie subgroup). Thus, there is reason to extend the theorem alluded to above to the case where M is an infinite dimensional Banach manifold.

Iain Raeburn considered this problem in [11]; for a certain kind of Banach submanifold M of a Banach space X , he defined a set $A_M \subset A \hat{\otimes} X$ and proved that $[A_M] \simeq [\Delta, M]$. However, his definition of A_M is not entirely satisfactory due to difficulties he had with applying the holomorphic functional calculus in an infinite dimensional setting.

We would like to describe an improved version of Raeburn's result and show how to use it to characterize $H^3(\Delta, Z)$. Our improvement rests on an infinite dimensional extension of the holomorphic functional calculus due to Waelbroeck [17]. We describe this in §1. In §2 we present our improvement of Raeburn's result. Sections 3, 4, and 5 are devoted to applications. Section 3 relates a certain category of projective Banach A-modules to the category of Banach space bundles over Δ. Section 3 relates a category of "twisted tensor products" of A with a Banach albegra B to the category of Banach algebra bundles over Δ with fiber B. In §5 we point out that if B has certain properties (which are possessed by the algebra of operators on Hilbert space), then isomorphism classes of Banach algebra bundles over Δ with fiber B form a group isomorphic to $H^3(\Delta, Z)$. Thus, $H^3(\Delta, Z)$ can be characterized as the group of isomorphism classes of "twisted products" of A with B. The same sort of analysis, but with B replaced by a finite dimensional matrix algebra, results in an isomorphism between the Brauer group of A and the torsion subgroup of $H^3(\Delta, Z)$.

The purpose of this paper is to simply announce some of our results along these lines. Proofs will appear in another paper. There are some mysteries connected with this subject which will be pointed out as we come to them. We hope to be able to clarify some of these before writing up the details.

§1. An infinite dimensional functional calculus.

Let A be a commutative Banach algebra with identity. The holomorphic functional calculus for A specifies a way of assigning to each n-tuple $\alpha = (a_1, \ldots, a_n)$ of elements of A and to each function f holomorphic on a neighborhood of the joint spectrum of α, an element $f(\alpha) \in A$. This assignment satisfies certain continuity and composition properties and, in the case where f is a polynomial, is obtained by simply substituting $a_1,\ldots,a_n$ for $z_1,\ldots,z_n$ (cf. [6]).

The purpose of this section is to state a theorem which re-casts and generalizes the standard holomorphic functional calculus. The theorem is essentially due to Waelbroeck [17] although he states it differently.

Let X be a complex Banach space and let A be as above. We denote the completed projective tensor product of A with X by $A \hat{\otimes} X$. This is a Banach A-module under the action of A on $A \hat{\otimes} X$ described on elementary tensors by $a(b \otimes x) = ab \otimes x$. In general, by a free Banach A-module we mean a module of this form.

If Δ is the maximal ideal space of A and $a \to \hat{a} : A \to C(\Delta)$ the Gelfand transform, then for each $\alpha = \Sigma a_i \otimes x_i \in A \hat{\otimes} X$ we have an element $\hat{\alpha} = \Sigma \hat{a}_i x_i \in C(\Delta, X)$ -- the space of continuous maps from Δ to X. We call $\hat{\alpha}$ the Gelfand transform of α and $\sigma(\alpha) = \hat{\alpha}(\Delta) \subset X$ the spectrum of α. Note that $\sigma(\alpha)$ is a compact subset of X.

If $C(\Delta,X)$ is given the sup norm, then $\alpha \to \hat{\alpha} : A \hat{\otimes} X \to C(\Delta,X)$ is norm decreasing. It follows that if U is open in X then

$$A_U = \{\alpha \in A \hat{\otimes} X : \sigma(\alpha) \subset U\}$$

is open in $A \hat{\otimes} X$. An open set in a Banach space will be called a Banach domain. Thus, to each Banach domain U we have assigned a domain A_U in a free Banach A-module.

If $U \subset X$ and $V \subset Y$ are Banach domains then a holomorphic map $f : U \to V$ is a continuous map with a complex linear Frechet derivative $f'(x) \in L(X,Y)$ at each $x \in U$ (cf. [9]).

A special type of holomorphic map $f : X \to Y$ is one of the form $f(x) = g(x,\ldots,x)$, where $g : X^n \to Y$ is a bounded symmetric multi-linear map. In fact, these are exactly the holomorphic maps f which are n-homogeneous in the sense that $f(zx) = z^n f(x)$ for $z \in \mathbb{C}$. Every holomorphic map between Banach domains has local series expansions in homogeneous terms (cf. [9]).

If $f(x) = g(x,\dots,x)$ is n-homogeneous and $\alpha = \Sigma a_i \otimes x_i \in A \hat{\otimes} X$, then one can define an element $A_f(\alpha) \in A \hat{\otimes} Y$ by

$$A_f(\alpha) = \sum_{i_1,\dots,i_n} a_{i_1} \dots a_{i_n} \otimes g(x_{i_1},\dots,x_{i_n}) .$$

That is, $A_f : A \hat{\otimes} X \to A \hat{\otimes} Y$ is constructed by applying the A-multilinear map

$(a_1 \otimes x_1) \times \dots \times (a_n \otimes x_n) \to a_1 \dots a_n \otimes f(x_1,\dots,x_n) : (A \hat{\otimes} X)^n \to A \hat{\otimes} Y$

to the diagonal in $(A \hat{\otimes} X)^n$.

The map A_f is a holomorphic map from $A \hat{\otimes} X$ to $A \hat{\otimes} Y$ which has the property that its Frechet derivative $f'(\alpha) : A \hat{\otimes} X \to A \hat{\otimes} Y$ is an A-module homomorphism for each $\alpha \in A \hat{\otimes} X$. We call a holomorphic map between domains in Banach A-modules which has this property an A-map.

Waelbroeck's version of the holomorphic functional calculus says that the correspondence $f \to A_f$ described above can be extended to other holomorphic maps between Banach domains.

<u>THEOREM</u> 1. There is a unique way of assigning to each holomorphic map $f : U \to V$ between Banach domains an A-map $A_f : A_U \to A_V$ such that:

(a) if f is n-homogeneous then A_f is defined as above;

(b) if $f : U \to V$ and $g : V \to W$ are holomorphic, then $A_g \circ A_f = A_{g \circ f}$;

(c) if $\varphi : A \to B$ is an identity preserving homomorphism between commutative Banach algebras and $f : U \to V$ is holomorphic, then $(\varphi \otimes 1) \circ A_f = B_f \circ (\varphi \otimes 1)$;

(d) for fixed $\alpha \in A_U$, $f \to A_f(\alpha)$ is continuous in the topology of uniform convergence on neighborhoods of $\sigma(\alpha)$.

If we consider only domains U and V in finite dimensional Banach spaces then this theorem reduces to the standard holomorphic functional calculus.

In the language of category theory, conclusion (b) of the theorem means that A determines a covariant functor $A_* = \{U \to A_U, f \to A_f\}$ from the category of Banach domains and holomorphic maps to the category of Banach A-module domains and A-maps. Conclusion (c) means that an algebra homomorphism $A \to B$ determines a natural transformation of functors $A_* \to B_*$.

Now a complex Banach manifold is a manifold modeled on complex Banach domains, with holomorphic maps as transition functions. Thus, it is very tempting to try to extend the functor A_* to the category of complex Banach manifolds and holomorphic maps. Unfortunately, we do not now have a completely general and satisfactory way of doing this, but we give some partial results in this direction in the next section.

§2. Banach Manifolds.

We shall consider only Banach manifolds which are embedded nicely in a Banach space. Specifically, we shall assume throughout this section that X is a complex Banach space, U is a domain in X, M is a complex Banach manifold, and M is embedded as a closed sub-manifold of U in such a way that its complex tangent space $T(M)_x$ is a complemented subspace of X for each $x \in M$.

We wish to associate a set $A_M \subset A_U$ to the manifold M. One is tempted to set $A_M = \{\alpha \in A \hat{\otimes} X : \sigma(\alpha) \subset M\}$. However, even if M is a complemented linear subspace of X (say $X = M \oplus N$), this is the wrong definition. If A is not semi-simple we would have $(A \hat{\otimes} M) \oplus (\mathrm{Rad}(A) \hat{\otimes} N) \subset A_M$, whereas the "correct" definition of A_M should yield $A_M = A \hat{\otimes} M$ in this case. Even if A is semi-simple there are difficulties with this definition if A fails to satisfy the approximation property for Banach spaces.

If M is a holomorphic retract of U (say $\pi : U \to M$ is the retraction) then there is no problem. We obtain a good definition by setting

$$A_M = \{\alpha \in A_U : A_\pi(\alpha) = \alpha\} .$$

It turns out that A_M as defined in this fashion is independent of the neighborhood U and the retraction π. Furthermore, the definition is functorial in the sense that a holomorphic map $f : M \to N$ induces a holomorphic map $A_f : A_M \to A_N$ whenever M and N are holomorphic retracts of Banach domains, and the usual composition law is satisfied by the assignment $f \to A_f$.

We have a definition for A_M when M is not a holomorphic neighborhood retract, but it is not yet clear that the definition is functorial. If we set

$$\widetilde{M} = \{(x,p) \in X \oplus L(X) : x \in M , p^2 = p, p(X) = T(M)_x\} ,$$

then $\widetilde{M}$ is a closed submanifold of a domain in $X \oplus L(X)$ and, in fact, is a neighborhood retract. Furthermore, the projection $\pi_1(x,p) = x$ from $\widetilde{M}$ to M makes $\widetilde{M}$ into a particularly nice kind of fiber bundle over M -- the fibers are contractible and are Banach complex homogeneous spaces. It follows from work of Raeburn [11] that such a bundle has what we shall call the lifting property:

> Given a (finite dimensional) Stein manifold Ω and a holomorphic map $f : \Omega \to M$, there is a holomorphic map $\widetilde{f} : \Omega \to \widetilde{M}$ such that $f = \pi_1 \circ \widetilde{f}$.

We now set

$$A_M = \{A_{\pi_1}(\alpha) \in A_U : \alpha \in A_{\widetilde{M}}\} .$$

That is, A_M is the set of all first elements of pairs in $(A \hat{\otimes} X) \oplus (A \hat{\otimes} L(X))$ belonging to $A_{\widetilde{M}}$.

In [11] Raeburn defines a set which we shall call A_M^0. This is the set of all $A_f(\alpha) \in A_U$, where $\alpha \in A \otimes \mathbb{C}^n$ for some n and f is a holomorphic map from a neighborhood of $\sigma(\alpha)$ into M. It follows from the lifting property for the bundle $\widetilde{M} \to M$ that $A_M^0 \subset A_M$. In fact, A_M^0 is dense in A_M.

In practice, M is usually given in the form

$$M = \{x \in U : f(x) = 0\}$$

where f is some simple holomorphic function from U to a Banach space Y. One would then like to be able to characterize A_M as the set of $\alpha \in A_U$ such that $A_f(\alpha) = 0$, or at least as a set determined solely in terms of f. A method which yields practical results is as follows: we assume that at each point $x \in M$ the map $f'(x) \in L(X,Y)$ has $T(M)_x$ as kernel and a complemented subspace of Y as image. Under these circumstances we say M is stably determined by f. We can then assert that $f'(x)$ has a pseudo-inverse for each $x \in M$, that is, there is a map $b \in L(Y,X)$ such that $\ker b$ is a complement for $\operatorname{im} f'(x)$, $\operatorname{im} b$ is a complement for $\ker f'(x)$ and b restricted to $\operatorname{im} f'(x)$ is the inverse of $f'(x)$ restricted to $\operatorname{im} b$.

If we set

$N = \{(x,b) \in X \times L(Y,X) : x \in M, b$ is a pseudo-inverse of $f'(x)\}$

then N is a holomorphic neighborhood retract in $X \times L(Y,X)$ and a holomorphic bundle over M with the lifting property described earlier. We now set

$$A_M^f = \{A_{\pi_1}(\alpha) : \alpha \in A_N\}$$

where $\pi_1 : X \times L(Y,X)$ is projection on the first factor. It turns out that A_M^f is the set of all $\beta \in A_U$ such that $A_f(\beta) = 0$ and $A_{f'}(\beta) \in A \hat{\otimes} L(X,Y)$ has a pseudo-inverse in $A \hat{\otimes} L(Y,X)$. Using the lifting property for the bundle $N \to M$ and the implicit function theorem we can prove:

<u>PROPOSITION 1.</u> If f stably determines M then $A_M^0 \subset A_M^f \subset A_M$, A_M^f is open in A_M and A_M^0 is dense in A_M. If M is finite dimensional then these three sets are the same and agree with $\{\alpha \in A_U : A_f(\alpha) = 0\}$.

To prove that $A_M^f = A_M = \{\alpha \in A_U : A_f(\alpha) = 0\}$ in general appears to require a lifting theorem for maps $\Omega \to M$ and bundles $N \to M$ in which the domain Ω is allowed to be infinite dimensional. We do not know whether or not such a lifting theorem is possible.

The set A_M is a sub-manifold of A_U with the property that its tangent space at each $\alpha \in A_M$ is a sub-module of $A \hat{\otimes} X$ of the form im p , where p is an idempotent in the algebra $A \hat{\otimes} L(X)$ (which acts naturally as an algebra of endomorphisms of $A \hat{\otimes} X$). Since A_M^f is open in A_M it is a manifold with the same property.

In general, we do not know whether or not A_M is closed in A_U or whether the correspondence $M \to A_M$ is functorial. That is, given manifolds $M \subset U$ and $N \subset V$ and a holomorphic map $f : M \to N$, does f induce a map $A_f : A_M \to A_N$? We can define $A_f : A_M \to A_V$ by setting $A_f(\alpha) = A_{f \circ \pi_1}(\beta)$ for $\beta \in A_{\tilde{M}}$ with $A_{\pi_1}(\beta) = \alpha$. Note, we could prove that $A_f(\alpha) \in A_N$ if we could lift $f \circ \pi_1 : \tilde{M} \to N$ to a map from $\tilde{M}$ to $\tilde{N}$. Also note that A_f does map A_M^0 into A_N^0 and, hence, A_M is mapped into the closure of A_N in A_V . Thus, if we could prove that A_N is always closed in A_V we could prove that $f \to A_f$ is functorial. For finite dimensional manifolds N , A_N is closed and there is no problem.

Although A_M^0 is not a sub-manifold of A_U , it is locally path connected. Since this is also true of A_M and A_M^f and since A_M^0 is dense in A_M , we have that $[A_M^0] = [A_M^f] = [A_M]$, where for any topological space S , $[S]$ denotes the set of components of S .

Raeburn [11] has proved that if M is a Banach complex homogeneous space (a quotient of a complex Banach Lie group by a closed complex Banach Lie subgroup) then the Gelfand transform $\alpha \to \hat{\alpha} : A \hat{\otimes} X \to C(\Delta, X)$ induces a bijection $[A_M^0] \to [\Delta, M]$ where $[\Delta, M] = [C(\Delta, M)]$ is the set of homotopy classes of maps from Δ to M . In view of our remarks above, we have the following extension of Raeburn's Theorem:

THEOREM 2. If each component of M is a Banach complex homogeneous space, then the induced maps $[A_M^0] \to [A_M] \to [\Delta, M]$ are bijections. If f stably determines M then $[A_M^f] \to [\Delta, M]$ is also a bijection.

§3. Regular Projective Modules.

If B is a (possibly non-commutative) Banach algebra, then $A \hat{\otimes} B$ is also a Banach algebra (with multiplication defined by $(a_1 \otimes b_1) \cdot (a_2 \otimes b_2) = a_1 a_2 \otimes b_1 b_2$ on elementary tensors). If M is the set of idempotents in B then M is a closed Banach submanifold of an open set U in B and, in fact, if we choose $U = \{b \in B : \sigma(b) \subset \{|z| < 1/2\} \cup \{|z-1| < 1/2\}\}$ then M is a holomorphic retract of U (cf. [11]). It turns out that A_M is the set of all idempotents of $A \hat{\otimes} B$.

Now each component of M is a Banach complex homogeneous space and, hence, by Theorem 2 we have that $[A_M] \to [\Delta, M]$ is a bijection. Furthermore, the components of A_M can be characterized as the orbits of the action of the identity component of $(A \hat{\otimes} B)^{-1}$ on M given by $(\beta, p) \to \beta p \beta^{-1}$. A detailed discussion of all this appears in Raeburn [11].

Now we consider the case where $B = L(X)$ for some Banach space X. An idempotent $p \in A \hat{\otimes} L(X)$ determines an idempotent module homomorphism $p : A \hat{\otimes} X \to A \hat{\otimes} X$ and, hence, a module direct sum composition $A \hat{\otimes} X = P \oplus Q$, where $P = \operatorname{im} p$ and $Q = \ker p$. However, not every module direct sum decomposition of $A \hat{\otimes} X$ arises in this way, because not every module projection $A \hat{\otimes} X \to A \hat{\otimes} X$ comes from an element of $A \hat{\otimes} L(X)$.

In general, a module homomorphism $A \hat{\otimes} X \to A \hat{\otimes} Y$ of the form $a \otimes x \to \sum ab_i \otimes \ell_i(x)$ for $\varphi = \sum b_i \otimes \ell_i \in A \hat{\otimes} L(X,Y)$ will be called a regular homomorphism. Thus, the projections on $A \hat{\otimes} X$ which come from idempotents of $A \hat{\otimes} L(X)$ are the regular projections. The decomposition $A \hat{\otimes} X = P \oplus Q$ determined by a regular projection will be called a regular decomposition and the summands P, Q will be called regular summands of $A \hat{\otimes} X$.

If $\varphi : A \hat{\otimes} X \to A \hat{\otimes} Y$ is any module homomorphism, then we define its Gelfand transform $\hat{\varphi} : \Delta \to L(X,Y)$ by $\hat{\varphi}(\delta)x = \varphi(1 \otimes x)^{\wedge}(\delta)$. Since $\varphi(1 \otimes x) \in A \hat{\otimes} Y$ its Gelfand transform $\varphi(1 \otimes x)^{\wedge}$ is in $C(\Delta, Y)$. It follows that $\hat{\varphi} : \Delta \to L(X,Y)$ is continuous in the strong operator topology on $L(X,Y)$. In general, this is the most that can be said. However, if φ is regular, then $\hat{\varphi}$ is actually norm continuous. In particular, if $p : A \hat{\otimes} X \to A \hat{\otimes} X$ is a module projection then $\hat{p} : \Delta \to L(X)$ is projection valued. The set of all $(\delta, \hat{p}(\delta)x) \in \Delta \times X$ is a bundle over Δ with Banach space fibers, but since $\hat{p}$ is only strongly continuous, this bundle may not be locally trivial. However, if p is a regular projection then $\hat{p}$ is norm continuous and the bundle is locally trivial.

With this observation in mind, we construct a category associated to A which the Gelfand transform sends to the category of locally trivial Banach space bundles over Δ . An object in this category will be a regular projection on $A \hat{\otimes} X$ for some Banach space X . Given regular projections p on $A \hat{\otimes} X$ and q on $A \hat{\otimes} Y$, a morphism $\varphi : p \to q$ will be a regular homomorphism $\varphi : A \hat{\otimes} X \to A \hat{\otimes} Y$ such that $\varphi p = \varphi = q\varphi$. For each p , the identity morphism from p to p is p itself. Thus, for a morphism $\varphi : p \to q$, an inverse is a morphism $\psi : q \to p$ with $\varphi \circ \psi = p$ and $\psi \circ \varphi = q$.

One may think of an object p in this category as an A-module (the image P of p) together with a specific representation of P as a summand in a regular decomposition of a free module. A morphism can be thought of as a module homomorphism $\varphi : P \to Q$ whose trivial extension to a map $A \hat{\otimes} X \to A \hat{\otimes} Y$ is regular. One must be careful here: it is conceivable that the same module could have two representations as regular summands of free modules which yield non-isomorphic elements of our category.

Direct summands of free modules are called projective modules. An object in our category is a certain kind of projective module together with some additional structure determined by the embedding. We will call such objects regular projective A-modules. Morphisms between such objects will be called regular homomorphisms.

As noted above, the Gelfand transform associates to each idempotent in $A \hat{\otimes} L(X)$ a locally trivial Banach space bundle over Δ. We declare a homomorphism between two such bundles E and F to be a norm continuous section of the bundle $\mathrm{Hom}(E,F)$. Then a regular homomorphism between regular projective A-modules induces (via the Gelfand transform) a morphism between the associated bundles. Thus, if $R\text{-}\mathrm{Proj}(A)$ is the category of regular projective A-modules and $B\text{-}\mathrm{Bun}(\Delta)$ is the category of locally trivial Banach space bundles over Δ, then the Gelfand transform induces a functor $R\text{-}\mathrm{Proj}(A) \to B\text{-}\mathrm{Bun}(\Delta)$.

With a little more work, Theorem 2 when applied to the manifold of idempotents in a Banach space yields the following theorem which is essentially due to Raeburn [11]:

THEOREM 3. Each object in $B\text{-}\mathrm{Bun}(\Delta)$ is isomorphic to one in the image of $R\text{-}\mathrm{Proj}(A)$. Furthermore, two objects in $R\text{-}\mathrm{Proj}(A)$ are isomorphic if and only if their images in $B\text{-}\mathrm{Bun}(\Delta)$ are isomorphic.

This generalizes a theorem of Novodvorskii [10] which says the analogous thing for the category $\mathrm{Proj}(A)$ of finitely generated projective A-modules and the category $\mathrm{Vect}(\Delta)$ of finite dimensional complex vector bundles over Δ.

The category $R\text{-}\mathrm{Proj}(A)$ is still somewhat mysterious. To clear up the mystery one needs an intrinsic characterization of the additional structure imposed on an A-module by an embedding of it as a regular direct summand of a free module.

Note that if $M \subset X$ is a sub-manifold satisfying the conditions of section 1, then $A_M \subset A \hat{\otimes} X$ is a sub-manifold whose tangent space at each point is a regular direct summand of $A \hat{\otimes} X$ and, hence, determines an object in R-Proj(A) .

§4. Twisted Products of Algebras.

One can quite easily make sense of the notions of direct sum and tensor product in the category R-Proj(A) . Once tensor products are defined, we define a regular A-algebra to be a regular projective A-module P together with a regular homomorphism $m : P \hat{\otimes} P \to P$ which satisfies the usual associative law. The functor R-Proj(A) → B-Bun(Δ) of the previous section transforms such a pair (P,m) into a locally trivial bundle of Banach algebras over Δ . Under certain additional conditions we can prove an analogue of Theorem 3 for this correspondence between regular A-algebras and Banach algebra bundles. We describe the extra conditions below.

Let B be a Banach algebra with identity and let $L^n(B,B)$ denote the Banach space of all n-linear maps from B^n to B . We define a map $f : L^2(B,B) \to L^3(B,B)$ by

$$f(m)(b_1,b_2,b_3) = m(m(b_1,b_2),b_3) - m(b_1,m(b_2,b_3)) .$$

Then f is holomorphic and $f^{-1}(0)$ is exactly the set of all associative multiplications on B . Now the multiplications on B which are equivalent to the given one comprise the image of the holomorphic map $g : L(B)^{-1} \to L^2(B,B)$ defined by

$$g(\varphi)(b_1,b_2) = \varphi^{-1}(\varphi(b_1)\varphi(b_2)) .$$

Note g(1) is the given multiplication $(b_1,b_2) \to b_1b_2$, the derivative g'(1) of g at 1 is the map $\delta^1 : L(B) \to L^2(B,B)$ defined by

$$\delta^1\varphi(b_1,b_2) = b_1\varphi(b_2) - \varphi(b_1b_2) + \varphi(b_1)b_2 ,$$

and the derivative of f at g(1) is the map $\delta^2 : L^2(B,B) \to L^3(B,B)$ defined by

$$\delta^2\psi(b_1,b_2,b_3) = b_1\psi(b_2,b_3) - \psi(b_1b_2,b_3) + \psi(b_1,b_2b_3) - \psi(b_1,b_2)b_3 .$$

These are the coboundary maps of degree one and two, respectively, in the Hochschild complex for B with coefficients in B (cf. [12]). In particular $\ker \delta^2/\mathrm{im}\, \delta^1$ is the second Hochschild cohomology group $H^2(B,B)$.

We will say that B is a stable Banach algebra if $H^2(B,B) = 0$ and δ^2 has closed complemented image and kernel (equivalently, $H^2(B,B) = 0$ and δ^2 has a pseudo-inverse). This assumption has the following implications:

PROPOSITION 2. Let B be a stable Banach algebra and let $M = \mathrm{im}\, g$ be the set of associative multiplications on B equivalent to the given one. Then:

(a) M is open in $f^{-1}(0)$;

(b) there is a domain $U \subset L^2(B,B)$ such that M is a closed sub-manifold of U ;

(c) M is stably determined by f on U.

(d) M is a Banach complex homogeneous space.

Clearly the space M is one to which we can apply Theorem 2. However, for the theorem we are seeking we must work with a somewhat more complicated space N. For a Banach space X, we consider the space of pairs $(p,m) \in L(X) \oplus L^2(X,X)$, where p is an idempotent which projects on a subspace Y isomorphic to B as a Banach space, and where m restricted to $Y \times Y$ is a multiplication isomorphic to the given one on B and m is zero on $\ker p \times Y$ and $Y \times \ker p$. The space N is then a sub-manifold of a domain in $L(X) \times L^2(X,X)$, is a Banach complex homogeneous space, and is stably determined by a holomorphic function h related to f.

If we apply Theorem 2 to the manifold N (for appropriate choices of X) and interpret the set A_N^h, we obtain Theorem 4 described below.

For a regular A-algebra P one can define a Hochschild complex by replacing the space $L^n(B,B)$ of multilinear maps by the space of regular homomorphisms from $\overset{n}{\otimes} P$ to P. For each n, the resulting object can be interpreted as a regular projective A-module. The coboundary maps δ^n are defined in more or less the usual way and yield regular homomorphisms. If $\ker \delta^2 = \mathrm{im}\, \delta^1$ and δ^2 has a regular pseudo-inverse we say that P is a stable regular A-algebra. This forces the fibers of the corresponding bundle of algebras on Δ to be stable Banach algebras. If these fibers are all isomorphic to a fixed stable algebra B, then we say that P is a stable twisted product of A with B. Our theorem is then:

THEOREM 4. Let B be a stable Banach algebra. Then the set of isomorphism classes of twisted tensor products of A with B is in bijective correspondence with the set of isomorphism classes of locally trivial Banach algebra bundles over Δ with fiber B.

We are not completely happy with the definition of stable twisted products of A with B. A cleaner definition may be possible which yields a class of algebras for which Theorem 4 remains valid. Part of the problem here is the lack of answers to some of the questions raised in §2.

If B is a finite dimensional algebra all the difficulties with infinite dimensional holomorphy, topological tensor products, and norm verses strong operator topologies disappear. Stable twisted products of A with B are then easily described: they are A-algebras P which are finitely generated and projective as A-modules and which have the property that for each maximal ideal $I \subset A$ the factor algebra P/IP is isomorphic to B. The condition that B be stable is simply that $H^2(B,B) = 0$. This condition is satisfied, in particular, by direct sums of matrix algebras.

§5. Third Cech Cohomology of Δ.

Let B be a Banach algebra with identity. The inner automorphisms of B are those of the form φ_u, where $u \in B^{-1}$ and

$\varphi_u(b) = ubu^{-1}$. The map $u \to \varphi_u$ is a holomorphic map and a group homomorphism of B^{-1} into Aut(B) . If B has trivial center and every automorphism of B is inner then we have an exact sequence of Banach Lie groups

$$0 \to \mathbb{C}^* \to B^{-1} \to \mathrm{Aut}(B) \to 0$$

where $\mathbb{C}^* = \mathbb{C}\setminus\{0\}$. In particular, this expresses B^{-1} as a holomorphic fiber bundle over Aut(B) with fiber $\mathbb{C}^*$.

If B^{-1} is contractible then Aut(B) will have vanishing homotopy groups except in degree 2 , where the homotopy groups will be Z . This follows from the homotopy long exact sequence for fiber bundles and the analogous statement for $\mathbb{C}^*$ (with 2 replaced by 1). It follows that a classifying space for bundles with structure group Aut(B) must have vanishing homotopy groups except in degree 3 , where the group will be Z . That is, a classifying space for Aut(B) must be an Eilenberg-MacLane space K(Z,3) (cf. [15]) and, hence, also a classifying space for third Čech cohomology.

Since locally trivial Banach algebra bundles over Δ with fiber B are bundles with fiber B and structure group Aut(B) , we conclude that the set of isomorphism classes of such bundles forms a group isomorphic to the third Čech group $H^3(\Delta, Z)$. In view of Theorem 4, we have:

<u>THEOREM</u> 5. Let B be a stable Banach algebra such that

(1) B has trivial center;

(2) all automorphisms of B are inner;

(3) B^{-1} is contractible.

Then the set of isomorphism classes of stable twisted products of A with B forms a group (under tensor product) which is isomorphic to $H^3(\Delta, Z)$.

The algebra $B = L(H)$ for H a Hilbert space satisfies the

conditions of the theorem provided H is infinite dimensional.

If we apply Theorem 4 in the case where B is a finite dimensional matrix algebra, we obtain a characterization of the Brauer group of A . The Brauer group is constructed from the set of isomorphism classes of twisted products of A with matric algebras. This set is a semigroup under tensor product. The Brauer group is a factor group of the Grothendieck group of this semigroup. The subgroup that is factored out is the one generated by A-algebras of the form $End_A(P)$ where P is a finitely generated projective A-module. The following theorem generalizes Serre's characterization [7] of the Brauer group of C(X) :

<u>THEOREM</u> 6. The Brauer group of A is isomorphic to the torsion subgroup of $H^3(\Delta, Z)$.

We should mention that our Theorem 4 bears a striking resemblance to a theorem of Dixmier and Douady [3] which classifies C^*-algebras formed from certain bundles of operator algebras. A mysterious thing here is that $H^3(\Delta, Z)$ still does the classifying even though the bundles considered by Dixmier and Douady are defined using the strong operator (rather than the norm) topology.

REFERENCES

1. Arens, R., The group of invertible elements of a commutative Banach algebra, Studia Math. (1963), 21-23.

2. _________, To what extent does the space of maximal ideals determine the algebra?, <u>Function Algebras</u>, Birtel, Scott Foresman, Chicago, 1966.

3. Dixmier, J. and Douady, A., Champs continus d'espaces Hilbertiens et C*-algebras, Bull. Soc. Mat. France, 91 (1963), 227-284.

4. Eidlin, V. L., The topological characteristics of the space of maximal ideals of a Banach algebra, Vestnik Leningrad Univ., 22 (1967), 173-174.

5. Forster, O., Functiontheoretische Hilfsmittel im der theorie der kommutativen Banach-algebren (in manuscript).

6. Gamelin, T. W., Uniform Algebras, Prentice-Hall, Englewood Cliffs, N. J., 1969.

7. Grothendieck, A., Le group de Brauer I: algébras d'azumaya et interprétations diverses, Séminaire Bourbaki, 1964/65, Expose 290.

8. Husemöller, D., Fiber bundles, McGraw-Hill, New York, 1966.

9. Nachbin, L., Topology on spaces of holomorphic mappings, Erg. Math Band 47, Springer-Verlag, Berlin, 1969.

10. Novodvorskii, M. E., Certain homotopical invariants of spaces of maximal ideals, Mat. Zametki, 1 (1967), 487-494.

11. Raeburn, I., The relationship between a commutative Banach algebra and its maximal ideal space (to appear).

12. Raeburn, I., and Taylor, J. L., Hochschild cohomology and perturbations of Banach algebras (to appear).

13. Roydon, H. L., Function algebras, Bull. Amer. Math. Soc., 69 (1963), 281-298.

14. Shilov, G. E., On decomposition of a commutative normed ring in a direct sum of ideals, Math. Sb., 32 (1953), 353-364; Amer. Math. Soc. Transl. (2) 1 (1955), 37-48.

15. Spanier, E. H., Algebraic Topology, McGraw-Hill, New York, 1966.

16. Taylor, J. L., Topological invariants of the maximal ideal space of a Banach algebra, Adv. in Math. 19 (1976), 149-206.

17. Waelbroeck, L., Topological vector spaces and algebras, Springer-Verlag Lecture Notes in Math., No. 230, Springer-Verlag, Berlin, 1971.

H-COBORDISMS, PSEUDO-ISOTOPIES, AND ANALYTIC TORSION

J.B. Wagoner, University of California at Berkeley*

To provide a background for the study of "higher analytic torsion" this expository article discusses the geometry of h-cobordisms and pseudo-isotopies as it shows up in the algebraic K-theory groups K_1 and K_2. The program is to see whether there are invariants of the higher homotopy groups of the space of pseudo-isotopies which can be expressed analytically in the spirit of the Ray-Singer proposal for analytic Reidemeister-Franz torsion [24].

§1. H-cobordisms and K_1.

This section deals with the existence of product structures on an h-cobordism. The next section is about uniqueness of product structures.

An h-cobordisms between two smooth, closed, n-dimensional manifolds M and M' is an (n+1)-dimensional smooth manifold W with $\partial W = M \cup M'$ such that the inclusions $M \subset W$ and $M' \subset W$ are homotopy equivalences. In proving the Poincare conjecture in high dimensions Smale showed [26] that if $\pi_1 W = 0$ and $n \geq 5$ then there is a product structure on W. More precisely there is a diffeomorphism $f: M \times I \longrightarrow W$ satisfying $f(M \times 0) = M$, $f|M \times 0 =$ identity, and $f(M \times 1) = M$. Smale's result was generalized by Barden [1], Mazur [20], and Stallings [27] who showed how the Whitehead group $Wh(\pi_1 W)$ and Whitehead's theory of simple homotopy types provided the obstruction to getting a product structure in the non-simply connected case. This is the s-cobordism theorem and we shall outline its proof. A good survey article is [21].

First recall the definition of the Whitehead group. For any associative ring with identity A let $GL(A) = \lim_{\vec{n}} GL_n(A)$ where $GL_n(A)$ is the group of $n \times n$ invertible matrices over A. Let E(A) be the subgroup

*Supported in part by the National Science Foundation.

of GL(A) generated by elementary matrices $e_{ij}(\lambda)$ for $i \neq j$ and $\lambda \in A$ where $e_{ij}(\lambda)$ is the identity on the diagonal and has λ in the (i,j) spot as the only non-zero entry off the diagonal. Then $E(A) = [GL(A), GL(A)]$ by the Whitehead Lemma [21, Lemma 1.1] and K_1 is defined to be the abelian group

$$K_1(A) = GL(A)/E(A).$$

If π is a group and $Z[\pi]$ is its integral group ring, let

$$Wh(\pi) = K_1(Z[\pi])/\langle \pm g \rangle$$

where $\langle \pm g \rangle$ is the subgroup generated by matrices

$$\begin{pmatrix} \pm g & & & & \\ & 1 & & & 0 \\ & & 1 & & \\ & 0 & & 1 & \\ & & & & \ddots \end{pmatrix}$$

for $g \in \pi$.

Now let M and M' be compact n-manifolds with (possibly empty) boundaries ∂M and $\partial M'$. A relative h-cobordim between M and M' is a smooth (n+1)-manifold W with $\partial W = M \cup M' \cup U$ where U is an h-cobordism between ∂M and $\partial M'$ equipped with a product structure $g\colon (\partial M \times I; \partial M \times 0, \partial M \times 1) \longrightarrow (U; \partial M, \partial M')$ and where both $M \subset W$ and $M' \subset W$ are homotopy equivalences as in the diagram

(1)

$$\begin{array}{ccccc} \partial M & \text{———} & U & \text{———} & \partial M' \\ M \{ & & W & & \} M' \\ \partial M & \text{———} & U & \text{———} & \partial M' \end{array}$$

<u>The s-Cobordism Theorem</u>. <u>Suppose</u> $n \geq 5$ <u>and let</u> $\pi = \pi_1 M \sim \pi_1 W$. <u>Then there is an element</u> $\tau = \tau(W;M,M')$ <u>of</u> $Wh(\pi)$ <u>which vanishes iff the product structure on</u> U <u>extends to a product structure</u>

$$G\colon (M \times I, M \times 0, M \times 1) \to (W; M, M').$$

Moreover every element of $Wh(\pi)$ *arises as the torsion of some relative h-cobordisms on* M.

This theorem also holds in the setting of piecewise-linear and topological manifolds. See [16], [19]. In the language of simple homotopy types W is a product iff both the inclusions $M \subset W$ and $M' \subset W$ are simple homotopy equivalences. A theory of higher simple homotopy types has been developed by A. Hatcher in [11]. While in practice one often takes the case where ∂M and $\partial M'$ are empty, the relative version with non-empty boundaries is quite useful. See §3 below. Some references for the proof of this theorem are [22] and [17]. For information on the Whitehead group and simple homotopy theory see [2], [7], [18], [21].

In discussing the proof we shall assume for simplicity that ∂M and $\partial M'$ are empty. The idea is to try to find a smooth function $f: W \to I$ with no critical points satisfying $f^{-1}(0) = M$ and $f^{-1}(1) = M'$. Then the integral curves of the gradient of f computed with respect to a choice of Riemannian metric on W give rise to a product structure:

(2)

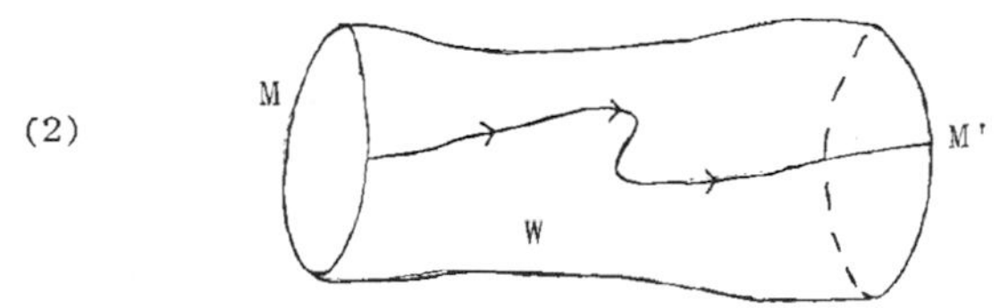

Since any two metrics can be joined by a path, the product structures arising from f in this way are all isotopic.

So begin with

Step 1. Definition of τ.

For $n \geq 5$ it is always possible to find a Morse function $f:W \to I$ with isolated, non-degenerate critical points satisfying (a) $f^{-1}(0) = M$, $f^{-1}(1) = M'$, and f has critical points only of index λ and $\lambda+1$ where $2 \leq \lambda \leq n-2$, (b) $f(p) > \frac{1}{2} > f(q)$ whenever p and q are critical points of index $\lambda + 1$ and λ respectively.

Let $W_\lambda = f^{-1}([0,\frac{1}{2}])$, $\partial_- W_\lambda = M$, $W_{\lambda+1} = f^{-1}([\frac{1}{2},1])$, and $\partial_- W_{\lambda+1} = f^{-1}(\frac{1}{2})$ as in the diagram

(3)

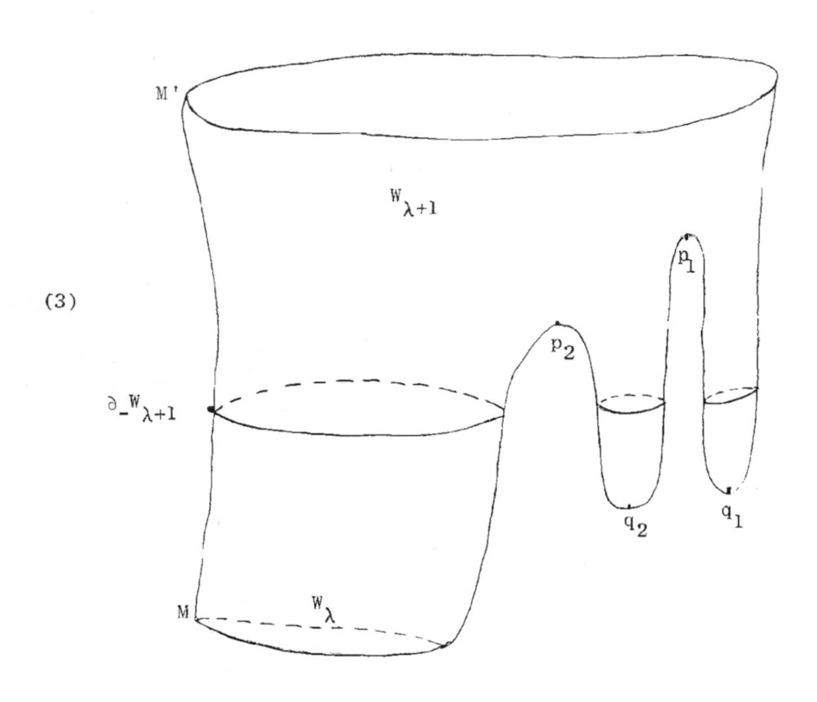

Let ρ: $\tilde{W} \to W$ be the universal covering and for $Y \subset W$ let $\tilde{Y} = \rho^{-1}(Y)$. The chain complex

$$\partial: C_{\lambda+1} \to C_\lambda$$

where $C_\mu = H_\mu(\tilde{W}_\mu, \partial_- \tilde{W}_\mu)$ has the same homology as $H_*(\tilde{W},\tilde{M}) = 0$ and hence ∂ is an isomorphism over $Z[\pi]$. Each C_μ is a free module over $Z[\pi]$ with rank m equal to the number of critical points of index μ, which is the same for $\mu = \lambda, \lambda + 1$. This boundary operator gives rise to an invertible $m \times m$ matrix as follows: Choose a gradient-like vector field η for f in general position. This means first that $df_x(\eta_x) > 0$ for x not a critical point of f and that when p is a critical point of index μ there is a local coordinate system $(x_1,\ldots,x_{n+1})$ about p so that

$$f(x_1,\ldots,x_{n+1}) = f(p) - x_1^2 - \ldots - x_\mu^2 + x_{\mu+1}^2 + \ldots + x_{n+1}^2$$

and

$$\eta(x_1,\ldots,x_{n+1}) = (-2x_1,\ldots,-2x_\mu, 2x_{\mu+1},\ldots,2x_{n+1})$$

Second, letting the stable manifold $W(p)$ of p be defined by $W(p) = \{x \in W \mid \lim_{t\to\infty} \varphi_t(x) = p\}$ where φ_t is the one-parameter group of diffeomorphisms generated by η, we require that $W(p)$ and $W(q)$ have empty intersection when index p = index q. In terms of handlebody theory this means that we have built up W from M by first attaching λ-handles simultaneously and then adding all the $(\lambda+1)$-handles. Note that $W(p) \cap W_\mu$ is a μ-disc with boundary $W(p) \cap \partial_- W_\mu$ a $(\mu-1)$-sphere. Now choose an ordering $p_1,\ldots,p_n$ and $q_1,\ldots,q_n$ of the critical points of index $\lambda+1$ and λ respectively. Choose orientations for the stable manifolds and liftings $\tilde{W}(p_i)$ and $\tilde{W}(q_j)$ in the universal cover. The discs $\tilde{W}(p_i) \cap \tilde{W}_{\lambda+1}$ give a basis e_i, $1 \leq i \leq n$, for $C_{\lambda+1}$ and the $\tilde{W}(q_j) \cap \tilde{W}_\lambda$ give a basis for C_λ. With respect to these bases ∂ becomes an invertible $m \times m$ matrix $\partial(\eta,f)$ which depends on the choice of η. The torsion $\tau = \tau(W;M,M')$ is defined to be

$$[\partial(\eta,f)] \in Wh(\pi)$$

Step 2. Vanishing of τ.

Suppose that $\tau = 0$. By introducing pairs of critical points which cancel each other such as p_1 and q_1 in diagram (3) and by making an appropriate choice of ordering of the critical points and of orientations and liftings of their stable manifolds, we can assume that

$$(*) \qquad \partial(\eta,f) = \prod_{\alpha=1}^{r} e_{i_\alpha j_\alpha}(\lambda_\alpha)$$

in $GL_m(Z[\pi])$ for m large. To get a function on W with no critical points we need two lemmas.

Handle Addition Lemma. (See [17, Lemma 5] and [22: 4.7, 7.6]). *Suppose p_i and p_j have index $\lambda + 1$ where $3 \leq \lambda + 1 \leq n-1$ and let $f(p_i) > f(p_j)$. Let $\gamma \in Z[\pi]$. Then there is a deformation (η_t, f) of (η, f) where η is gradient-like for f, $0 \leq t \leq 1$, such that (η_1, f) is in general position and the basis for $C_{\lambda+1}$ determined by the new stable manifolds is*

$$e_1, \ldots, e_i + \gamma \cdot e_j, \ldots, e_j, \ldots, e_n.$$

In other words, η can be deformed to η_1 so that $\partial(\eta_1, f) = e_{ij}(\gamma) \cdot \partial(\eta, f)$.

Rearrangement Lemma. (see [22, Th. 4.1]). *Given critical points p and q of index $\lambda + 1$ with $f(p) > f(q)$ there is a deformation (η_t, f_t) of (η, f) where η_t is gradient-like and in general position for f_t with each f_t satisfying (a) and (b) such that $f_1(p) < f_1(q)$.*

This lemma is illustrated by the diagram

(4) 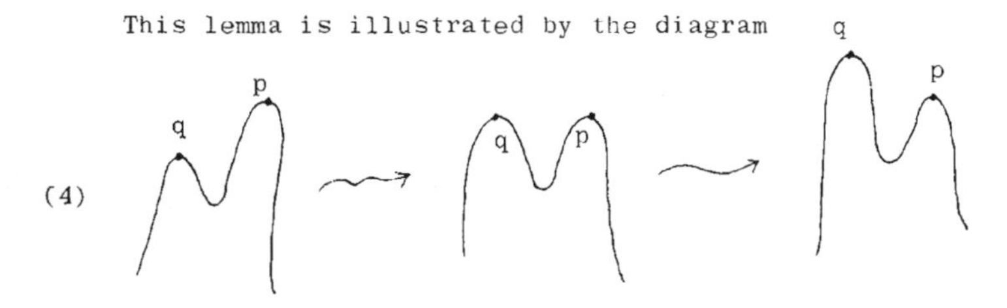

It follows that corresponding to a particular choice of a product decomposition (*) there is a deformation (η_t, f_t) of (η, f) to a pair (η_1, f_1) in general position such that $\partial(\eta_1, f_1)$ = identity. The pairs of critical points are now cancelled using the Smale-Whitney process as in the deformation

(5)

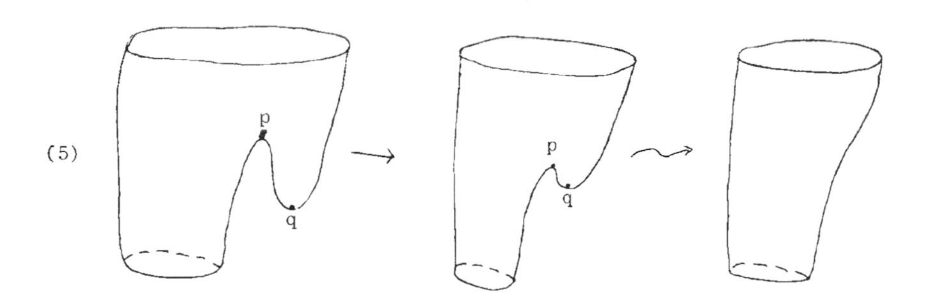

to give a function g with no critical points on W. See [22: 5.4,6.4], [17].

Suppose we are given another product decomposition

(**) $$\partial(\eta,f) = \prod_{\beta=1}^{s} e_{i_\beta j_\beta}(\lambda_\beta)$$

This gives another deformation (η'_t, f'_t) of (η, f) to (η'_1, f'_1) where $\partial(\eta'_1, f'_1)$ = id and then cancelling pairs of critical points gives $g':W \to I$ with no critical points. Algebraically we have

(***) $$\prod_{\beta=1}^{s} e_{i_\beta j_\beta}(\lambda_\beta) \cdot \prod_{\alpha=r}^{1} e_{i_\alpha j_\alpha}(-\lambda_\alpha) = \text{id}$$

Thus the difference between g and g' is measured, in part, by "homotopy classes" of product decompositions of the identity matrix. This is made precise in §2 using the algebraic K-theory group K_2. The other part of the difference between g and g' comes from the uniqueness question for Smale-Whitney cancellation.

Incidentally, the proof that $\partial(\eta,f)$ gives a well-defined class in $Wh(\pi)$ is to join two such pairs (η,f) and (η',f') in general position by a generic path (η_t,f_t) and to show that as in the handle addition lemma one gets from $\partial(\eta,f)$ to $\partial(\eta',f')$ by multiplying on the left (and on the right) by elementary matrices. The indeterminacy in $\partial(\eta,f)$ coming from the ordering of critical points, choice of orientations, etc., is taken care of by $\langle \pm g \rangle$.

§2 Pseudo-isotopies and K_2.

Let $f,g: (M \times I; M \times 0, M \times 1) \to (W;M,M')$ be two product structures on the h-cobordisms W. Then f and g are isotopic rel $M \times 0$ iff $g^{-1}f: M \times I \to M \times I$ is isotopic to the identity rel $M \times 0$. This means that uniqueness of product structures is measured by $\pi_0\mathcal{P}(M)$ where $\mathcal{P}(M)$ is the space of pseudo-isotopies of M; that is, the space of diffeomorphisms $h: M \times I \to M \times I$ such that $h|M \times 0$ = identity. For simplicity we assume ∂M is empty. The main result in the relative case is exactly the same for the space $\mathcal{P}(M,\partial M)$ of pseudo-isotopies f of $M \times I$ with $f|\partial M \times I$ = identity also. J. Cerf initiated work on this problem by showing that $\pi_0\mathcal{P}(M) = 0$ whenever $\pi_1 M = 0$ and dim $M \geq 5$. See [5] and [6]. As an application we have $\pi_0 \text{ Diff}^+(D^n) = \pi_0\mathcal{P}(S^{n-1}) = 0$ for $n \geq 6$ where Diff^+ is the space of orientation preserving diffeomorphisms of the n-disc D^n.

Cerf's approach relates $\mathcal{P}(M)$ to the space $\mathcal{E}$ implicitly considered in §1 of all smooth functions $f: M \times I \to I$ with no critical points satisfying $f^{-1}(0) = M \times 0$ and $f^{-1}(1) = M \times 1$. Let $p: M \times I \to I$ be the standard projection $p(x,t) = t$. The map $\theta: \mathcal{P}(M) \to \mathcal{E}$ taking f to $p \circ f^{-1}$ is a Serre fibration with $\theta^{-1}(p) = \mathcal{J}$, the space of isotopies of M starting at the identity. Recall that an isotopy $f: M \times I \to M \times I$ is a pseudo-isotopy which preserves levels: $f(M \times t) = M \times t$. $\mathcal{J}$ is contractible since it is just the space of all paths in Diff M starting at the identity. Hence

$$\mathcal{P}(M) \to \mathcal{E}$$

is a homotopy equivalence. This means that $\pi_0\mathcal{P} = \pi_0\mathcal{E}$ and so uniqueness of product structures is equivalent to measuring the difference of two functions on $M \times I$ with no critical points as indicated in §1. Note that since $\mathcal{J}$ is contractible, a pseudo-isotopy is isotopic to the identity of $M \times I$ iff it is isotopic to an isotopy of the identity on M. Let $\mathcal{F}$ be the space of all smooth functions $f: M \times I \to I$ with $f^{-1}(0) = M \times 0$ and $f^{-1}(1) = M \times 1$. Since $\mathcal{F}$ is contractible, $\pi_1(\mathcal{F},\mathcal{E}) =$

$\pi_0(\mathcal{S})$. The idea then for measuring the difference between $f,g \in \mathcal{S}$ is to connect them by a path f_t in $\mathcal{J}$ and to try to deform this path down into $\mathcal{S}$. To each generic path one can associate, as the discussion of §1 indicates, the "homotopy class" of a product decomposition (***) of the identity, and this gives the K_2 obstruction.

Recall the definition of Wh_2. Let A be any associative ring with identity. Define the Steinberg group St(A) as the group generated by $x_{ij}(\lambda)$ where $1 \leq i \neq j < \infty$ and $\lambda \in A$ modulo the relations

(a) $\quad x_{ij}(\lambda)\cdot x_{ij}(\mu) = x_{ij}(\lambda+\mu)$

(b) $\quad [x_{ij}(\lambda), x_{k\ell}(\mu)] = 1 \quad$ for $\quad i \neq \ell \quad$ and $j \neq k$

(c) $\quad [x_{ij}(\lambda), x_{jk}(\mu)] = x_{ik}(\lambda\mu) \quad$ for $\quad i,j,k$ distinct.

There is a natural homomorphism $St(A) \to E(A)$ sending $x_{ij}(\lambda)$ to $e_{ij}(\lambda)$ and Milnor [23] defines the abelian group $K_2(A)$ as the kernel of the central extension

$$0 \to K_2(A) \to St(A) \to E(A) \to 1.$$

Let $A = Z[\pi]$ for $\pi = \pi_1 M$. Let $W(\pm\pi)$ be the subgroup of St(A) generated by the words $w_{ij}(\pm g) = x_{ij}(\pm g)\cdot x_{ji}(\mp g^{-1}) \cdot x_{ij}(\pm g)$ for $f \in \pi$. Define

$$Wh_2(\pi) = K_2(Z[\pi]) \bmod \{W(\pm\pi) \cap K_2(Z[\pi])\}.$$

To get the second algebraic obstruction for pseudo-isotopies let $(Z_2 \times \pi_2)[\pi_1]$ denote the abelian group of all functions $\pi_1 \to Z_2 \times \pi_2$ which are zero except on finitely many elements of π_1. Here $\pi_1 = \pi_1 M$ and $\pi_2 = \pi_2 M$. Any element of $(Z_2 \times \pi_2)[\pi_1]$ is written as a finite formal sum $\sum_i \alpha_i \cdot \sigma_i$ where $\alpha_i \in Z_2 \times \pi_2$ and $\sigma_i \in \pi_1$. Let π_1 act trivially on Z_2 and as usual on π_2. If $\alpha \in Z_2 \times \pi_2$ and $\tau \in \pi_1$, denote the action by α^τ. Define

$$Wh_1^+(\pi_1; Z_2 \times \pi_2)$$

to be $(Z_2 \times \pi_2)[\pi_1]$ modulo the subgroup generated by $\alpha\cdot\sigma - \alpha^\tau\cdot\tau\sigma\tau^{-1}$ and $\beta\cdot 1$ for $\alpha, \beta \in Z_2 \times \pi_2$ and $\sigma, \tau \in \pi_1$. Here 1 is the identity of π_1.

Theorem.[1] For any compact, connected, smooth manifold M with dim M $\geqslant$ 6 there is an isomorphism

$$\pi_0\mathscr{P}(M) \simeq Wh_2(\pi_1) \oplus Wh_1^+(\pi_1; Z_2 \times \pi_2)$$

The first obstruction was worked out by the author, and the full theorem was done concurrently and independently by Hatcher and also by Volodin [28]. For a complete proof see [15]. Shorter expositions are in [13], [14], and [30]. Igusa has improved the dimension condition to dim M $\geq$, and the result has been extended to the piecewise-linear and topological categories in [4]. Hatcher [12] and Volodin [28] have obtained some results on $\pi_1\mathscr{P}$. Notice that when $\pi_1 M = 0$, $Wh_2(\pi_1)$ vanishes because $K_2(Z) = Z_2$ is generated by $w_{12}(1)^4$. See [23, §10]. Also $Wh_1^+(1; Z_2 \times \pi_2)$ clearly vanishes and so we get Cerf's result in the simply connected case when n $\geq$ 6.

Here are some algebraic facts about $Wh_2(\pi)$:

(i) It is probably true that Wh_2(finite group) is finite. [8], [10]

(ii) $Wh_2(Z_{20})$ has at least order five (Milnor)

(iii) $Wh_2(\pi \times Z) = Wh_2(\pi) \oplus Wh(\pi) \oplus$? See [29].

It is conjectured that ? = $Nil_1^+(G) \oplus Nil_1^-(G)$.

(iv) $Wh_2(G) = 0$ if G is free or free abelian. (Swan and Gersten using methods of Quillen.)

For more information about K_2 see [9].

§3. Non-trivial pseudo-isotopies from $Wh(\pi)$.

In this section we discuss Siebenmann's construction [25] of a map

$$\rho\colon \pi_0\mathscr{P}(M \times S^1) \to Wh(\pi_1 M)$$

which is surjective. This gave the first examples of non-trivial pseudo-isotopies before the connection of $\pi_0\mathscr{P}$ with K_2 was firmly established. The algebraic analogue is (iii) of §2, and in §4 the idea

(1) See important footnote following the references.

that $Wh(\pi) \subset Wh_2(\pi \times Z)$ will be used to give evidence for the existence of an "analytic formula" for the Wh_2 obstruction.

Let $f: M \times S^1 \times I \to M \times S^1 \times I$ be a pseudo-isotopy and let $\tilde{f}: M \times R \times I \to M \times R \times I$ be the unique lifting of f to the infinite cyclic cover $M \times R \times I$ of $M \times S^1 \times I$. Choose a $\delta > 0$ large enough that $\tilde{f}(M \times [\delta, \infty) \times I)$ lies in the interior of $M \times [0, \infty) \times I$ as in

(6)

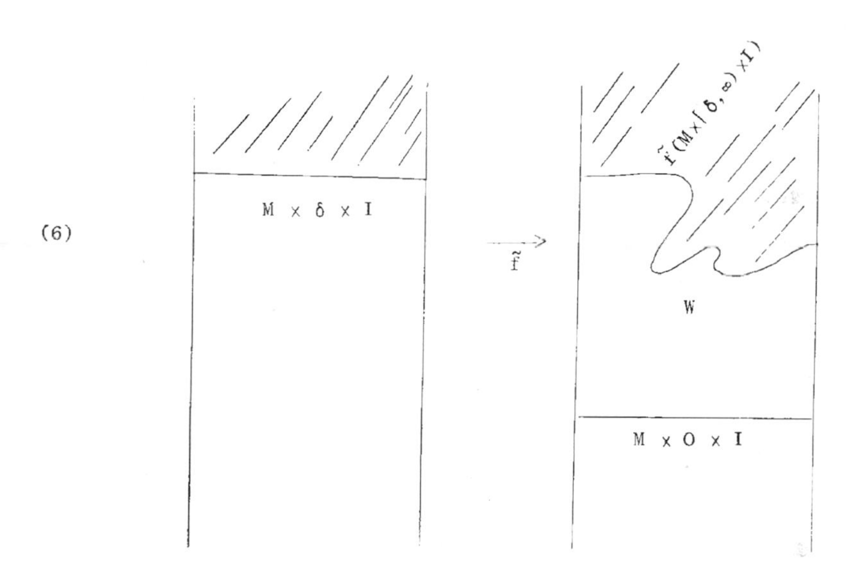

Let W be the closure of the complement of $\tilde{f}(M \times [\delta, \infty) \times I)$ in $M \times [0, \infty) \times I$. This is an h-cobordism on $M \times [0, \delta] \times 0$ with a product structure $\partial(M \times [0, \delta]) \times I \cong M \times 0 \times I \cup \tilde{f}(M \times \delta \times I)$ given by the identity on $M \times 0 \times I$ and by $\tilde{f}: M \times \delta \times I \to \tilde{f}(M \times \delta \times I)$ on $M \times \delta \times I$. Define

$$\rho(f) = \tau(W) \in Wh(\pi_1 M).$$

This only depends on the isotopy class of f and gives the desired map.

To see ρ is onto let $\tau \in Wh(\pi_1 M)$. Construct a relative h-cobordism W_τ on $M \times [0, \frac{1}{2}]$ with torsion τ and similarly get $W_{-\tau}$ on $M \times [\frac{1}{2}, 1]$ with torsion $-\tau$ as in the diagram

$M \times [\frac{1}{2},1]$ $W_{-\tau}$

(7)

$M \times [0,\frac{1}{2}]$ W_{τ}

By the Addition Theorem [21, 7.4], the relative h-cobordism $W = W_{\tau} \cup W_{-\tau}$ on $M \times [0,1]$ has torsion $\tau + (-\tau) = 0$ and so there is a product structure $g: W \to M \times [0,1] \times I$ which is the identity on $M \times 0 \times I$ and $M \times 1 \times I$. In other words $M \times [0,1] \times I$ is the union of two relative h-cobordisms $W'_{\tau} = g(W_{\tau})$ and $W'_{-\tau} = g(W_{-\tau})$ as in

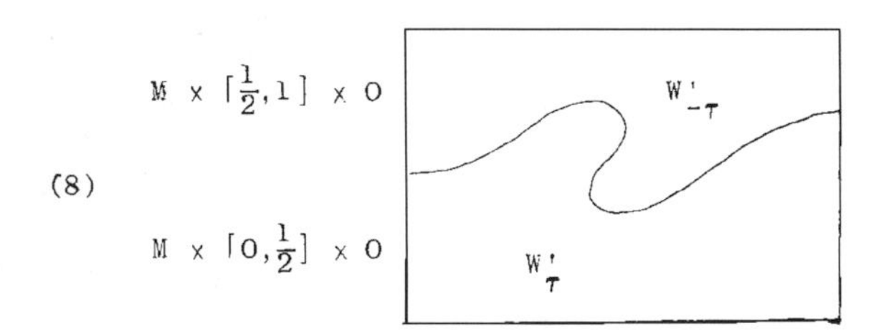

Similarly let $h: M \times [\frac{1}{2},\frac{3}{2}] \times I \to W'_{-\tau} \cup W'_{\tau}$ be a relative product structure such that $h|M \times \alpha \times I = g$ for $\alpha = \frac{1}{2},\frac{3}{2}$ as in

(9)

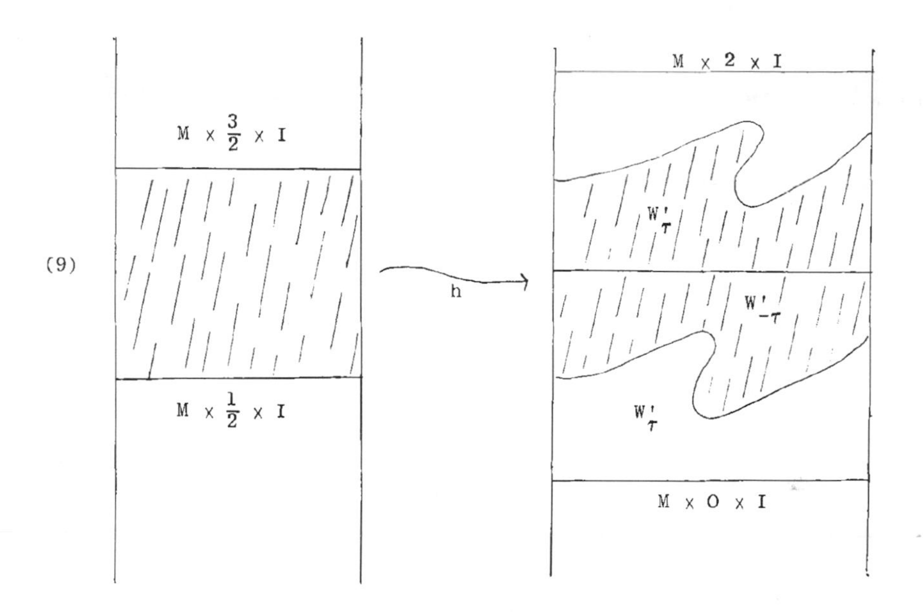

By translating along the R coordinate extend h to a diffeomorphism of $M \times R \times I$ to itself which is equivariant with respect to the Z action and which is therefore the lifting of a pseudo-isotopy f on $M \times S^1 \times I$. Clearly from the construction we have $\rho(f)$ = torsion of $W'_\tau = \tau$. This shows ρ is surjective.

§4. Higher analytic torsion.

Any homomorphism $\chi : \pi \to U_n$ induces a ring homomorphism from $Z[\pi]$ into the $n \times n$ matrices $M_n(C)$ over the complex numbers C and so there is a homomorphism of algebraic K-theory groups

$$\chi_* : K_i(Z[\pi]) \to K_i(M_n(C)) \simeq K_i(C)$$

When $i = 1$, we get a homomorphism $\chi_* : Wh(\pi) \to K_1(C)/S^1 \simeq R^*/\langle\pm 1\rangle$ which gives rise to the corresponding Reidemeister-Franz R-torsion of an h-cobordism or of a closed manifold with χ acyclic. In [24] Ray and Singer proposed an analytic R-torsion in terms of the zeta function of the Laplacian acting on the deRham complex. Is there also an analytic formula for the $Wh_2(\pi)$ invariant of $\pi_0 \mathcal{P}$? Or more generally for that part of $\pi_i \mathcal{P}$ captured by an appropriately defined group $Wh_{i+2}(\pi)$

via a mapping induced by χ into a suitable quotient of $K_{i+2}(C)$? The example in this section indicates that one might expect an analytic interpretation for Wh_2 despite the fact that there is no continuous Steinberg symbol on $C^* \times C^*$.

For a pseudo-isotopy $f: M \times I \to M \times I$ the general approach might go like this: Choose a Riemannian metric $\mathcal{g}$ on $M \times I$ and let $\mathcal{g}'$ be the metric induced by f. Let Δ and Δ' be the corresponding Laplacians on the deRham complex with coefficients in χ. Then Δ and Δ' are unitarily equivalent, have the same eigenvalues, etc. Join $\mathcal{g}$ to $\mathcal{g}'$ by a generic path $\mathcal{g}_t$ of metrics and let Δ_t be the resulting one-parameter family of Laplacians. As t varies the eigenvalues and eigenspaces change and the question is how to associate to this data a K_2 type invariant.

In the case of pseudo-isotopy f on $M \times S^1 \times I$ we can be more precise about a candidate for the analytic version of the projection $Wh_2(\pi \times Z) \to Wh(\pi)$. Choose a Riemannian metric $\mathcal{g}$ on $M \times S^1 \times I$ which is a product near the boundary and lift this in the natural way to a metric $\mathcal{g}$ on the infinite cyclic cover $M \times R \times I$. Deform f so that it is the identity near $M \times S^1 \times 0$ and is of the form $(f|M \times S^1 \times 1) \times id$ near $M \times S^1 \times 1$. Let W be the relative h-cobordism as in Diagram (6) of §3. Put a metric μ on W by first letting $\mu = \mathcal{g}$ near $(M \times [0,\delta] \times 0) \cup (M \times 0 \times I)$ and μ = metric induced by $\tilde{f}$ from $\mathcal{g}$ near $\tilde{f}(M \times \delta \times I)$. Then define μ on the rest of W so that near the part of ∂W lying in $M \times R \times 1$ it is a product. In this setting of a manifold W corners it should be possible following [24] to define the analytic torsion $\log T_W(\chi)$ and to prove that $\log(T_W(\chi)/T_W(\varepsilon))$ is independent of how μ was filled in on W. Here $\chi: \pi_1 M \to U_n$ comes by restriction from the given representation $\chi: \pi_1(M \times S^1) \to U_n$ and $\varepsilon: \pi_1(M \times S^1) \to U_n$ is the trivial representation. Then we should have

$$\log \chi_*(\rho(f)) = \log(T_W(\chi)/T_W(\varepsilon)).$$

Finally, we give an example of a representation $\chi: G \to S^1$ where $\chi_*: K_2(Z[G]) \to K_2(C)$ is non-trivial. Let $G = \pi \times Z$ where π is the cyclic group of order five generated by u and Z is generated by t. Let $\chi: G \to S^1$ be given by $\chi(u) = \xi = e^{2\pi i/5}$ and $\chi(t) = x \in S^1$ where x is an element of a transcendence base X of C over Q, i.e., C is an algebraic extension of the transcendental extension Q(X) of Q.

<u>Claim</u>. $\chi_*(\{u + u^{-1} -1, t\}) \neq 0$

where $u + u^{-1} -1$ is the well-known [21, 6.6] generator of the infinite cyclic group $Wh(\pi)$ which is embedded in $Wh_2(\pi \times Z)$ by the correspondence $M \to \{M, t\}$ as in [29]. Note that $\chi_*: Wh(\pi) \to R^*/\langle \pm 1 \rangle$ takes $u + u^{-1} -1$ to $\xi + \xi^{-1} -1 = 2 \cos (2\pi/5) -1$. Following [23, 11.10] let $X' = X - \{x\}$. Then viewing Q(X) as the field of rational functions in the variable x with coefficients in Q(X') we get a discrete valuation ν corresponding to the ideal $\langle x \rangle$. Embedding C in the algebraic closure of the completion of Q(X) for the $\langle x \rangle$-adic topology gives a valuation ω on C with $\omega(z) = \frac{\nu(\text{Norm } z)}{n}$ where z has degree n over the completion. Since ξ is algebraic over Q, Norm $(\xi + \xi^{-1} -1) \in Q$ and $\omega(\xi + \xi^{-1} -1) = 0$. Consider the extension $Q(X, \xi)$ of Q(X). The valuation ω is discrete on $Q(X, \xi)$ and the corresponding tame symbol [23, 11.5] takes $\chi_*(\{u + u^{-1} -1, t\}) = \{\xi + \xi^{-1} -1, x\} = \{2 \cos(2\pi/5) -1, x\}$ in $K_2(Q(X, \xi))$ to the element $2 \cos (2\pi/5) -1$ of infinite order in the residue field. Hence it persists in $K_2(C)$ by the argument of [23, 11.10].

REFERENCES

1. D. Barden, <u>Structure of manifolds</u>, Thesis, Cambridge University (1963).

2. H. Bass, <u>Algebraic K-Theory</u>, Benjamin, Inc., New York, 1968.

3. ______ and J. Tate, <u>The Milnor ring of a global field</u>, Algebraic K-Theory II, Springer-Verlag Lecture Notes No. 342, pp. 349-446.

4. D. Burghelia and R. Lashof, <u>The homotopy type of the space of diffeomorphisms I and II</u>, Trans. Amer. Math. Soc. 196 (1974), 1-50.

5. J. Cerf, C.R. Congres Int. Math., Moscow,1966, pp. 429-437.

6. _______, La stratification naturelle des espaces de fonctions differentiables reeles et la theoreme de la pseudo-isotopie, Pub. Math. I.H.E.S. No. 39 (1970).

7. R.K. Dennis, The computation of Whitehead groups, lecture notes, Universitat Bielefeld,1973.

8. __________, Stability for K_2, to appear in Proceedings of the Conference on Orders and Group Rings, Ohio State University at Columbus, May 1972, Springer-Verlag Lecture Notes.

9. __________, and M. Stein, The functor K_2: A survey of computations and problems, Algebraic K-Theory II, Springer-Verlag Lecture Notes No. 342, pp. 243-280.

10. H. Garland, A finiteness theorem for K_2 of a number field, Ann. of Math. 94 (1971), pp. 534-548.

11. A. Hatcher, Higher simple homotopy theory, Annals of Math. 102 (1975), 101-137.

12. __________, preprint, Princeton University.

13. __________, Parametrized h-cobordism theory, Ann. Inst. Fourier, Grenoble, 23, 2 (1973), 61-74.

14. __________, Pseudo-isotopy and K_2, Algebraic K-Theory II, Springer -Verlag Lecture Notes No. 342, pp. 328-335.

15. __________ and J. Wagoner, Pseudo-isotopies of compact manifolds, Asterisque No. 6 (1973), Societe Mathematique de France.

16. J.F.P. Hudson, Piecewise linear topology, Benjamin, Inc., New York, 1968.

17. M. Kervaire, Le Theoreme de Barden-Mazur-Stallings, Comm. Math. Helv. 40 (1965), 31-42.

18. __________, S. Maumary, and G. deRham, Torsion et type simple d'homotopie, Springer-Verlag Lecture Notes in Mathematics No. 48.

19. R. Kirby and L. Siebenmann, Essays on Topological Manifolds, Smoothings, and Triangulations, preprint, University of Paris (Orsay).

20. B. Mazur, Differential topology from the point of view of simple homotopy theory, Publ. Math. I.H.E.S., No. 15.

21. J. Milnor, Whitehead torsion, Bull. Amer. Math. Soc. 72 (1966), 358-426.

22. __________, Lectures on the h-cobordism theorem, Notes by L. Siebenmann and J. Sondow, Princeton Mathematical Notes, Princeton University Press, 1965.

23. __________, Introduction to algebraic K-theory, Annals Studies No. 72, Princeton 1971.

24. D.B. Ray and I.M. Singer, R-torsion and the Laplacian on

Riemannian manifolds, Advances in Mathematics 7 (1971), 145-210.

25. L. Siebenmann, Torsion invariants for pseudo-isotopies on closed manifolds, Notices Amer. Math. Soc. 14 (1967), 942.

26. S. Smale, On the structure manifolds, Amer. J. Math. 84 (1962), 387-399.

27. J. Stallings, On infinite processes leading to differentiability in the complement of a point, Differential and Combinatorial Topology (A symposium in honor of M. Morse), Princeton University Press, 245-254.

28. I.A. Volodin, Uspeki No. 5, 1972.

29. J. Wagoner, On K_2 of the Laurent polynomial ring, Amer. J. Math. 93 (1971), 123-138.

30. ________, Algebraic invariants for pseudo-isotopies, Proceedings of Liverpool Singularities Symposium II, Springer-Verlag Lecture Notes No. 209, 164-190.

Footnote concerning the pseudo-isotopy theorem: Recently Igusa pointed out a difficulty in the proof of Part II of [15] that the Wh_1^+ obstruction is well defined. Hatcher has since shown the proof remains valid under the additional assumption that the first k-invariant in $H^3(\pi_1 M; \pi_2 M)$ vanishes. The Wh_2 obstruction of Part I of [15] is well defined for any connected, compact, smooth manifold.

Vol. 399: Functional Analysis and its Applications. Proceedings 1973. Edited by H. G. Garnir, K. R. Unni and J. H. Williamson. II, 584 pages. 1974.

Vol. 400: A Crash Course on Kleinian Groups. Proceedings 1974. Edited by L. Bers and I. Kra. VII, 130 pages. 1974.

Vol. 401: M. F. Atiyah, Elliptic Operators and Compact Groups. V, 93 pages. 1974.

Vol. 402: M. Waldschmidt, Nombres Transcendants. VIII, 277 pages. 1974.

Vol. 403: Combinatorial Mathematics. Proceedings 1972. Edited by D. A. Holton. VIII, 148 pages. 1974.

Vol. 404: Théorie du Potentiel et Analyse Harmonique. Edité par J. Faraut. V, 245 pages. 1974.

Vol. 405: K. J. Devlin and H. Johnsbråten, The Souslin Problem. VIII, 132 pages. 1974.

Vol. 406: Graphs and Combinatorics. Proceedings 1973. Edited by R. A. Bari and F. Harary. VIII, 355 pages. 1974.

Vol. 407: P. Berthelot, Cohomologie Cristalline des Schémas de Caracteristique p > o. II, 604 pages. 1974.

Vol. 408: J. Wermer, Potential Theory. VIII, 146 pages. 1974.

Vol. 409: Fonctions de Plusieurs Variables Complexes, Séminaire François Norguet 1970–1973. XIII, 612 pages. 1974.

Vol. 410: Séminaire Pierre Lelong (Analyse) Année 1972–1973. VI, 181 pages. 1974.

Vol. 411: Hypergraph Seminar. Ohio State University, 1972. Edited by C. Berge and D. Ray-Chaudhuri. IX, 287 pages. 1974.

Vol. 412: Classification of Algebraic Varieties and Compact Complex Manifolds. Proceedings 1974. Edited by H. Popp. V, 333 pages. 1974.

Vol. 413: M. Bruneau, Variation Totale d'une Fonction. XIV, 332 pages. 1974.

Vol. 414: T. Kambayashi, M. Miyanishi and M. Takeuchi, Unipotent Algebraic Groups. VI, 165 pages. 1974.

Vol. 415: Ordinary and Partial Differential Equations. Proceedings 1974. XVII, 447 pages. 1974.

Vol. 416: M. E. Taylor, Pseudo Differential Operators. IV, 155 pages. 1974.

Vol. 417: H. H. Keller, Differential Calculus in Locally Convex Spaces. XVI, 131 pages. 1974.

Vol. 418: Localization in Group Theory and Homotopy Theory and Related Topics. Battelle Seattle 1974 Seminar. Edited by P. J. Hilton. VI, 172 pages 1974.

Vol. 419: Topics in Analysis. Proceedings 1970. Edited by O. E. Lehto, I. S. Louhivaara, and R. H. Nevanlinna. XIII, 392 pages. 1974.

Vol. 420: Category Seminar. Proceedings 1972/73. Edited by G. M. Kelly. VI, 375 pages. 1974.

Vol. 421: V. Poénaru, Groupes Discrets. VI, 216 pages. 1974.

Vol. 422: J.-M. Lemaire, Algèbres Connexes et Homologie des Espaces de Lacets. XIV, 133 pages. 1974.

Vol. 423: S. S. Abhyankar and A. M. Sathaye, Geometric Theory of Algebraic Space Curves. XIV, 302 pages. 1974.

Vol. 424: L. Weiss and J. Wolfowitz, Maximum Probability Estimators and Related Topics. V, 106 pages. 1974.

Vol. 425: P. R. Chernoff and J. E. Marsden, Properties of Infinite Dimensional Hamiltonian Systems. IV, 160 pages. 1974.

Vol. 426: M. L. Silverstein, Symmetric Markov Processes. X, 287 pages. 1974.

Vol. 427: H. Omori, Infinite Dimensional Lie Transformation Groups. XII, 149 pages. 1974.

Vol. 428: Algebraic and Geometrical Methods in Topology, Proceedings 1973. Edited by L. F. McAuley. XI, 280 pages. 1974.

Vol. 429: L. Cohn, Analytic Theory of the Harish-Chandra C-Function. III, 154 pages. 1974.

Vol. 430: Constructive and Computational Methods for Differen tial and Integral Equations. Proceedings 1974. Edited by D. L. Colton and R. P. Gilbert. VII, 476 pages. 1974.

Vol. 431: Séminaire Bourbaki – vol. 1973/74. Exposés 436–452. IV, 347 pages. 1975.

Vol. 432: R. P. Pflug, Holomorphiegebiete, pseudokonvexe Gebiete und das Levi-Problem. VI, 210 Seiten. 1975.

Vol. 433: W. G. Faris, Self-Adjoint Operators. VII, 115 pages. 1975.

Vol. 434: P. Brenner, V. Thomée, and L. B. Wahlbin, Besov Spaces and Applications to Difference Methods for Initial Value Problems. II, 154 pages. 1975.

Vol. 435: C. F. Dunkl and D. E. Ramirez, Representations of Commutative Semitopological Semigroups. VI, 181 pages. 1975.

Vol. 436: L. Auslander and R. Tolimieri, Abelian Harmonic Analysis, Theta Functions and Function Algebras on a Nilmanifold. V, 99 pages. 1975.

Vol. 437: D. W. Masser, Elliptic Functions and Transcendence. XIV, 143 pages. 1975.

Vol. 438: Geometric Topology. Proceedings 1974. Edited by L. C. Glaser and T. B. Rushing. X, 459 pages. 1975.

Vol. 439: K. Ueno, Classification Theory of Algebraic Varieties and Compact Complex Spaces. XIX, 278 pages. 1975

Vol. 440: R. K. Getoor, Markov Processes: Ray Processes and Right Processes. V, 118 pages. 1975.

Vol. 441: N. Jacobson, PI-Algebras. An Introduction. V, 115 pages. 1975.

Vol. 442: C. H. Wilcox, Scattering Theory for the d'Alembert Equation in Exterior Domains. III, 184 pages. 1975.

Vol. 443: M. Lazard, Commutative Formal Groups. II, 236 pages. 1975.

Vol. 444: F. van Oystaeyen, Prime Spectra in Non-Commutative Algebra. V, 128 pages. 1975.

Vol. 445: Model Theory and Topoi. Edited by F. W. Lawvere, C. Maurer, and G. C. Wraith. III, 354 pages. 1975.

Vol. 446: Partial Differential Equations and Related Topics. Proceedings 1974. Edited by J. A. Goldstein. IV, 389 pages. 1975.

Vol. 447: S. Toledo, Tableau Systems for First Order Number Theory and Certain Higher Order Theories. III, 339 pages. 1975.

Vol. 448: Spectral Theory and Differential Equations. Proceedings 1974. Edited by W. N. Everitt. XII, 321 pages. 1975.

Vol. 449: Hyperfunctions and Theoretical Physics. Proceedings 1973. Edited by F. Pham. IV, 218 pages. 1975.

Vol. 450: Algebra and Logic. Proceedings 1974. Edited by J. N. Crossley. VIII, 307 pages. 1975.

Vol. 451: Probabilistic Methods in Differential Equations. Proceedings 1974. Edited by M. A. Pinsky. VII, 162 pages. 1975.

Vol. 452: Combinatorial Mathematics III. Proceedings 1974. Edited by Anne Penfold Street and W. D. Wallis. IX, 233 pages. 1975.

Vol. 453: Logic Colloquium. Symposium on Logic Held at Boston, 1972–73. Edited by R. Parikh. IV, 251 pages. 1975.

Vol. 454: J. Hirschfeld and W. H. Wheeler, Forcing, Arithmetic, Division Rings. VII, 266 pages. 1975.

Vol. 455: H. Kraft, Kommutative algebraische Gruppen und Ringe. III, 163 Seiten. 1975.

Vol. 456: R. M. Fossum, P. A. Griffith, and I. Reiten, Trivial Extensions of Abelian Categories. Homological Algebra of Trivial Extensions of Abelian Categories with Applications to Ring Theory. XI, 122 pages. 1975.

Vol. 457: Fractional Calculus and Its Applications. Proceedings 1974. Edited by B. Ross. VI, 381 pages. 1975.

Vol. 458: P. Walters, Ergodic Theory – Introductory Lectures. VI, 198 pages. 1975.

Vol. 459: Fourier Integral Operators and Partial Differential Equations. Proceedings 1974. Edited by J. Chazarain. VI, 372 pages. 1975.

Vol. 460: O. Loos, Jordan Pairs. XVI, 218 pages. 1975.

Vol. 461: Computational Mechanics. Proceedings 1974. Edited by J. T. Oden. VII, 328 pages. 1975.

Vol. 462: P. Gérardin, Construction de Séries Discrètes p-adiques. »Sur les séries discrètes non ramifiées des groupes réductifs déployés p-adiques«. III, 180 pages. 1975.

Vol. 463: H.-H. Kuo, Gaussian Measures in Banach Spaces. VI, 224 pages. 1975.

Vol. 464: C. Rockland, Hypoellipticity and Eigenvalue Asymptotics. III, 171 pages. 1975.

Vol. 465: Séminaire de Probabilités IX. Proceedings 1973/74. Edité par P. A. Meyer. IV, 589 pages. 1975.

Vol. 466: Non-Commutative Harmonic Analysis. Proceedings 1974. Edited by J. Carmona, J. Dixmier and M. Vergne. VI, 231 pages. 1975.

Vol. 467: M. R. Essén, The Cos $\pi\lambda$ Theorem. With a paper by Christer Borell. VII, 112 pages. 1975.

Vol. 468: Dynamical Systems – Warwick 1974. Proceedings 1973/74. Edited by A. Manning. X, 405 pages. 1975.

Vol. 469: E. Binz, Continuous Convergence on C(X). IX, 140 pages. 1975.

Vol. 470: R. Bowen, Equilibrium States and the Ergodic Theory of Anosov Diffeomorphisms. III, 108 pages. 1975.

Vol. 471: R. S. Hamilton, Harmonic Maps of Manifolds with Boundary. III, 168 pages. 1975.

Vol. 472: Probability-Winter School. Proceedings 1975. Edited by Z. Ciesielski, K. Urbanik, and W. A. Woyczyński. VI, 283 pages. 1975.

Vol. 473: D. Burghelea, R. Lashof, and M. Rothenberg, Groups of Automorphisms of Manifolds. (with an appendix by E. Pedersen) VII, 156 pages. 1975.

Vol. 474: Séminaire Pierre Lelong (Analyse) Année 1973/74. Edité par P. Lelong. VI, 182 pages. 1975.

Vol. 475: Répartition Modulo 1. Actes du Colloque de Marseille-Luminy, 4 au 7 Juin 1974. Edité par G. Rauzy. V, 258 pages. 1975. 1975.

Vol. 476: Modular Functions of One Variable IV. Proceedings 1972. Edited by B. J. Birch and W. Kuyk. V, 151 pages. 1975.

Vol. 477: Optimization and Optimal Control. Proceedings 1974. Edited by R. Bulirsch, W. Oettli, and J. Stoer. VII, 294 pages. 1975.

Vol. 478: G. Schober, Univalent Functions – Selected Topics. V, 200 pages. 1975.

Vol. 479: S. D. Fisher and J. W. Jerome, Minimum Norm Extremals in Function Spaces. With Applications to Classical and Modern Analysis. VIII, 209 pages. 1975.

Vol. 480: X. M. Fernique, J. P. Conze et J. Gani, Ecole d'Eté de Probabilités de Saint-Flour IV–1974. Edité par P.-L. Hennequin. XI, 293 pages. 1975.

Vol. 481: M. de Guzmán, Differentiation of Integrals in R^n. XII, 226 pages. 1975.

Vol. 482: Fonctions de Plusieurs Variables Complexes II. Séminaire François Norguet 1974–1975. IX, 367 pages. 1975.

Vol. 483: R. D. M. Accola, Riemann Surfaces, Theta Functions, and Abelian Automorphisms Groups. III, 105 pages. 1975.

Vol. 484: Differential Topology and Geometry. Proceedings 1974. Edited by G. P. Joubert, R. P. Moussu, and R. H. Roussarie. IX, 287 pages. 1975.

Vol. 485: J. Diestel, Geometry of Banach Spaces – Selected Topics. XI, 282 pages. 1975.

Vol. 486: S. Stratila and D. Voiculescu, Representations of AF-Algebras and of the Group U (∞). IX, 169 pages. 1975.

Vol. 487: H. M. Reimann und T. Rychener, Funktionen beschränkter mittlerer Oszillation. VI, 141 Seiten. 1975.

Vol. 488: Representations of Algebras, Ottawa 1974. Proceedings 1974. Edited by V. Dlab and P. Gabriel. XII, 378 pages. 1975.

Vol. 489: J. Bair and R. Fourneau, Etude Géométrique des Espaces Vectoriels. Une Introduction. VII, 185 pages. 1975.

Vol. 490: The Geometry of Metric and Linear Spaces. Proceedings 1974. Edited by L. M. Kelly. X, 244 pages. 1975.

Vol. 491: K. A. Broughan, Invariants for Real-Generated Uniform Topological and Algebraic Categories. X, 197 pages. 1975.

Vol. 492: Infinitary Logic: In Memoriam Carol Karp. Edited by D. W. Kueker. VI, 206 pages. 1975.

Vol. 493: F. W. Kamber and P. Tondeur, Foliated Bundles and Characteristic Classes. XIII, 208 pages. 1975.

Vol. 494: A Cornea and G. Licea. Order and Potential Resolvent Families of Kernels. IV, 154 pages. 1975.

Vol. 495: A. Kerber, Representations of Permutation Groups II. V, 175 pages. 1975.

Vol. 496: L. H. Hodgkin and V. P. Snaith, Topics in K-Theory. Two Independent Contributions. III, 294 pages. 1975.

Vol. 497: Analyse Harmonique sur les Groupes de Lie. Proceedings 1973–75. Edité par P. Eymard et al. VI, 710 pages. 1975.

Vol. 498: Model Theory and Algebra. A Memorial Tribute to Abraham Robinson. Edited by D. H. Saracino and V. B. Weispfenning. X, 463 pages. 1975.

Vol. 499: Logic Conference, Kiel 1974. Proceedings. Edited by G. H. Müller, A. Oberschelp, and K. Potthoff. V, 651 pages 1975.

Vol. 500: Proof Theory Symposion, Kiel 1974. Proceedings. Edited by J. Diller and G. H. Müller. VIII, 383 pages. 1975.

Vol. 501: Spline Functions, Karlsruhe 1975. Proceedings. Edited by K. Böhmer, G. Meinardus, and W. Schempp. VI, 421 pages. 1976.

Vol. 502: János Galambos, Representations of Real Numbers by Infinite Series. VI, 146 pages. 1976.

Vol. 503: Applications of Methods of Functional Analysis to Problems in Mechanics. Proceedings 1975. Edited by P. Germain and B. Nayroles. XIX, 531 pages. 1976.

Vol. 504: S. Lang and H. F. Trotter, Frobenius Distributions in GL_2-Extensions. III, 274 pages. 1976.

Vol. 505: Advances in Complex Function Theory. Proceedings 1973/74. Edited by W. E. Kirwan and L. Zalcman. VIII, 203 pages. 1976.

Vol. 506: Numerical Analysis, Dundee 1975. Proceedings. Edited by G. A. Watson. X, 201 pages. 1976.

Vol. 507: M. C. Reed, Abstract Non-Linear Wave Equations. VI, 128 pages. 1976.

Vol. 508: E. Seneta, Regularly Varying Functions. V, 112 pages. 1976.

Vol. 509: D. E. Blair, Contact Manifolds in Riemannian Geometry. VI, 146 pages. 1976.

Vol. 510: V. Poènaru, Singularités C^∞ en Présence de Symétrie. V, 174 pages. 1976.

Vol. 511: Séminaire de Probabilités X. Proceedings 1974/75. Edité par P. A. Meyer. VI, 593 pages. 1976.

Vol. 512: Spaces of Analytic Functions, Kristiansand, Norway 1975. Proceedings. Edited by O. B. Bekken, B. K. Øksendal, and A. Stray. VIII, 204 pages. 1976.

Vol. 513: R. B. Warfield, Jr. Nilpotent Groups. VIII, 115 pages. 1976.

Vol. 514: Séminaire Bourbaki vol. 1974/75. Exposés 453 – 470. IV, 276 pages. 1976.

Vol. 515: Bäcklund Transformations. Nashville, Tennessee 1974. Proceedings. Edited by R. M. Miura. VIII, 295 pages. 1976.